Arduino for Secret Agents

Transform your tiny Arduino device into a secret agent gadget and build a range of espionage projects with this practical guide for hackers

Marco Schwartz

BIRMINGHAM - MUMBAI

Arduino for Secret Agents

First published: November 2015

Production reference: 1121115

Published by Packt Publishing Ltd.
Livery Place
35 Livery Street
Birmingham B3 2PB, UK.

ISBN 978-1-78398-608-8

www.packtpub.com

Credits

Author
Marco Schwartz

Reviewer
Roberto Gallea

Commissioning Editor
Julian Ursell

Acquisition Editors
Harsha Bharwani

Prachi Bisht

Content Development Editor
Pooja Mhapsekar

Technical Editor
Bharat Patil

Copy Editor
Vibha Shukla

Project Coordinator
Francina Pinto

Proofreader
Safis Editing

Indexer
Mariammal Chettiyar

Production Coordinator
Arvindkumar Gupta

Cover Work
Arvindkumar Gupta

About the Author

Marco Schwartz is an electrical engineer, an entrepreneur, and a blogger. He has a master's degree in electrical engineering and computer science from Supélec, France, and a master's degree in micro engineering from the Ecole Polytechnique Fédérale de Lausanne (EPFL) in Switzerland.

He has more than five years of experience working in the domain of electrical engineering. Marco's interests gravitate around electronics, home automation, the Arduino and Raspberry Pi platforms, open source hardware projects, and 3D printing.

He has several websites about Arduino, including the Open Home Automation website that is dedicated to building home automation systems using open source hardware.

Marco has written another book on home automation and Arduino, called *Home Automation With Arduino: Automate Your Home Using Open-source Hardware*. He has also written a book on how to build Internet of Things projects with Arduino, called *Internet of Things with the Arduino Yun, Packt Publishing*.

About the Reviewer

Roberto Gallea, PhD, is a computer science researcher since 2007. He was working at the University of Palermo, Italy. He is committed to investigating fields such as medical imaging, multimedia, and computer vision. In 2012, he started enhancing his academic and personal projects with the use of analog and digital electronics, with particular involvement in the open source hardware and software platform, Arduino. Besides academic interests, he also conducts personal projects that are aimed at producing handcrafted items embedding invisible electronics, such as musical instruments, furniture, and LED devices. He has also been collaborating with contemporary dance companies for digital scenic and costume design.

www.PacktPub.com

Support files, eBooks, discount offers, and more

For support files and downloads related to your book, please visit www.PacktPub.com.

Did you know that Packt offers eBook versions of every book published, with PDF and ePub files available? You can upgrade to the eBook version at www.PacktPub.com and as a print book customer, you are entitled to a discount on the eBook copy. Get in touch with us at service@packtpub.com for more details.

At www.PacktPub.com, you can also read a collection of free technical articles, sign up for a range of free newsletters and receive exclusive discounts and offers on Packt books and eBooks.

https://www2.packtpub.com/books/subscription/packtlib

Do you need instant solutions to your IT questions? PacktLib is Packt's online digital book library. Here, you can search, access, and read Packt's entire library of books.

Why subscribe?

- Fully searchable across every book published by Packt
- Copy and paste, print, and bookmark content
- On demand and accessible via a web browser

Free access for Packt account holders

If you have an account with Packt at www.PacktPub.com, you can use this to access PacktLib today and view 9 entirely free books. Simply use your login credentials for immediate access.

Table of Contents

Preface   iii

Chapter 1: A Simple Alarm System with Arduino   1

Hardware and software requirements   1
Hardware configuration   4
Configuring the alarm system   6
Testing the alarm system   10
Summary   12

Chapter 2: Creating a Spy Microphone   13

Hardware and software requirements   13
Using the SD card   15
Testing the microphone   20
Building the spy microphone   22
Recording on the SD card   25
Summary   27

Chapter 3: Building an EMF Bug Detector   29

Hardware and Software requirements   29
Hardware configuration   31
Testing the LCD screen   33
Building the EMF bug detector   34
Summary   39

Chapter 4: Access Control with a Fingerprint Sensor   41

Hardware and software requirements   41
Hardware configuration   43
Enrolling your fingerprint   44
Controlling access to the relay   49
Accessing secret data   52
Summary   57

Chapter 5: Opening a Lock with an SMS   59

Hardware and software requirements   60
Hardware configuration   62
Testing the FONA shield   65
Controlling the relay   70
Opening and closing the lock   73
Summary   75

Chapter 6: Building a Cloud Spy Camera   77

Hardware and software requirements   77
Hardware configuration   80
Setting up your Dropbox account   82
Setting up your Temboo account   84
Saving pictures to Dropbox   88
Live streaming from the spy camera   93
Summary   95

Chapter 7: Monitoring Secret Data from Anywhere   97

Hardware and software requirements   98
Hardware configuration   99
Sending data to dweet.io   101
Monitoring the device remotely   108
Creating automated e-mail alerts   112
Summary   113

Chapter 8: Creating a GPS Tracker with Arduino   115

Hardware and software requirements   115
Hardware configuration   117
Testing the location functions   118
Sending a GPS location by SMS   127
Building a GPS location tracker   129
Summary   132

Chapter 9: Building an Arduino Spy Robot   133

Hardware and software requirements   133
Hardware configuration   134
Setting up the motor control   141
Setting up live streaming   147
Setting up the interface   148
Testing the surveillance robot   151
Summary   152

Index   153

Preface

The Arduino platform makes it really easy to build electronics projects in various domains, such as home automation, Internet of Things, wearable technology, and even healthcare. It's also the ideal platform to build amazing projects for secret agents, which is what we are going to do in this book.

Using the power and simplicity of the Arduino platform, we are going to see how to build several projects that can be easily used by any aspiring secret agent. From audio recorders to GPS trackers, you will be able to make your own secret agent toolkit using the Arduino platform after reading this book.

What this book covers

Chapter 1, *A Simple Alarm System with Arduino*, is about building an alarm system that is based on the Arduino platform with a motion sensor and a visual alarm.

Chapter 2, *Creating a Spy Microphone*, is about making a secret recording system that can record the conversations and noises in a room.

Chapter 3, *Building an EMF Bug Detector*, is about creating a very useful device for any secret agent: a detector to check whether there are other secret agent devices in a room.

Chapter 4, *Access Control with a Fingerprint Sensor*, is about creating an access control system using your own fingerprint.

Chapter 5, *Opening a Lock with an SMS*, is about building a project where the secret agent can open a lock just by sending a text message to the Arduino device.

Chapter 6, *Building a Cloud Spy Camera*, is about making a spy camera that can be accessed from anywhere in the world and can record pictures in Dropbox when motion is detected.

Chapter 7, Monitoring Secret Data from Anywhere, is about learning how to secretly record any kind of data and how to log in this data on the Cloud.

Chapter 8, Creating a GPS Tracker with Arduino, is about creating one of the most useful devices for a secret agent: a GPS tracker that indicates its position on a map in real time.

Chapter 9, Building an Arduino Spy Robot, is about making a small surveillance robot that can spy on your behalf.

What you need for this book

In the entire book, we will be using the Arduino platform so you will definitely need the latest version of the Arduino IDE software.

We will be using a wide range of Arduino boards, shields, and hardware components. You will find all the details about these requirements in the relevant chapters.

Who this book is for

This book is intended for those who want to build exciting secret agent projects using the Arduino platform. For example, it is for those people who are already experienced in using the Arduino platform and want to extend their knowledge by building projects for secret agents. It is also for the people who want to learn about electronics and programming as Arduino is the perfect platform for that.

Conventions

In this book, you will find a number of text styles that distinguish between different kinds of information. Here are some examples of these styles and an explanation of their meaning.

Code words in text, database table names, folder names, filenames, file extensions, pathnames, dummy URLs, user input, and Twitter handles are shown as follows: "Also, if the `alarm_mode` is going back to false, we need to deactivate the alarm immediately."

A block of code is set as follows:

```
if (alarm_mode == false) {

    // No tone & LED off
    noTone(alarm_pin);
    digitalWrite(led_pin, LOW);
}
```

Any command-line input or output is written as follows:

```
mjpg_streamer -i "input_uvc.so -d /dev/video0 -r 640x480 -f 25" -o
"output_http.so -p 8080 -w /www/webcam" &
```

New terms and **important words** are shown in bold. Words that you see on the screen, for example, in menus or dialog boxes, appear in the text like this: "Now, inside the parameters of the app, there are two things you need: the **App key**, and the **App secret**."

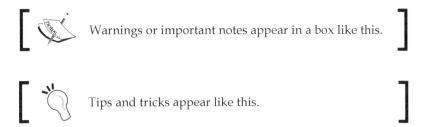

> Warnings or important notes appear in a box like this.

> Tips and tricks appear like this.

Reader feedback

Feedback from our readers is always welcome. Let us know what you think about this book—what you liked or disliked. Reader feedback is important for us as it helps us develop titles that you will really get the most out of.

To send us general feedback, simply e-mail feedback@packtpub.com, and mention the book's title in the subject of your message.

If there is a topic that you have expertise in and you are interested in either writing or contributing to a book, see our author guide at www.packtpub.com/authors.

Customer support

Now that you are the proud owner of a Packt book, we have a number of things to help you to get the most from your purchase.

Downloading the example code

You can download the example code files from your account at http://www.packtpub.com for all the Packt Publishing books you have purchased. If you purchased this book elsewhere, you can visit http://www.packtpub.com/support and register to have the files e-mailed directly to you.

Errata

Although we have taken every care to ensure the accuracy of our content, mistakes do happen. If you find a mistake in one of our books—maybe a mistake in the text or the code—we would be grateful if you could report this to us. By doing so, you can save other readers from frustration and help us improve subsequent versions of this book. If you find any errata, please report them by visiting http://www.packtpub.com/submit-errata, selecting your book, clicking on the **Errata Submission Form** link, and entering the details of your errata. Once your errata are verified, your submission will be accepted and the errata will be uploaded to our website or added to any list of existing errata under the Errata section of that title.

To view the previously submitted errata, go to https://www.packtpub.com/books/content/support and enter the name of the book in the search field. The required information will appear under the **Errata** section.

Piracy

Piracy of copyrighted material on the Internet is an ongoing problem across all media. At Packt, we take the protection of our copyright and licenses very seriously. If you come across any illegal copies of our works in any form on the Internet, please provide us with the location address or website name immediately so that we can pursue a remedy.

Please contact us at copyright@packtpub.com with a link to the suspected pirated material.

We appreciate your help in protecting our authors and our ability to bring you valuable content.

Questions

If you have a problem with any aspect of this book, you can contact us at questions@packtpub.com, and we will do our best to address the problem.

1
A Simple Alarm System with Arduino

I want to start this book with a simple project that any secret agent will want to have, a simple alarm system that will be activated whenever motion is detected by a sensor. This simple system is not only fun to make but will also help us to go over the basics of Arduino programming and electronics, which are the skills that we will use in this whole book.

It will basically be a simple alarm (a buzzer that makes sound, plus a red LED) combined with a motion detector. The user will also be able to stop the alarm by pressing a button.

We are going to do the following in this chapter:

- First, we are going to see what the requirements for this project are, in terms of hardware and software
- Then, we will see how to assemble the hardware parts for this project
- After that, we will configure our system using the Arduino IDE

Hardware and software requirements

First, let's see what the required components for this project are. As this is the first chapter of the book, we will spend a bit more time here to detail the different components, as these are components that we will be using in the whole book.

The first component that will be central to the project is the Arduino Uno board:

In several chapters of this book, this will be the 'brain' of the projects that we will make. In all the projects, I will be using the official Arduino Uno R3 board. However, you can use an equivalent board from another brand or another Arduino board, such as an Arduino Mega board.

Another crucial component of our alarm system will be the buzzer:

This is a very simple component that is used to make simple sounds with Arduino. You couldn't play an MP3 with it but it's just fine for an alarm system. You can, of course, use any buzzer that is available; the goal is to just make a sound.

After that, we are going to need a motion detector:

Here, I used a very simple PIR motion detector. This sensor will measure the infrared (IR) light that is emitted by moving objects in its field of view, for example, people moving around. It is really easy and quite cheap to interface with Arduino. You can use any brand that you want for this sensor; it just needs a voltage level of 5V in order to be compatible with the Arduino Uno board.

Finally, here is the list of all the components that we will use in this project:

- Arduino Uno (https://www.sparkfun.com/products/11021)
- Buzzer (https://www.sparkfun.com/products/7950)
- PIR (https://www.sparkfun.com/products/13285)
- LED (https://www.sparkfun.com/products/9590)
- 330 Ohm resistor (https://www.sparkfun.com/products/8377)
- Button (https://www.sparkfun.com/products/97)
- 1k Ohm resistor (https://www.sparkfun.com/products/8980)
- Breadboard (https://www.sparkfun.com/products/12002)
- Jumper wires (https://www.sparkfun.com/products/8431)

On the software side, the only thing that we will need in the first chapter is the latest version of the Arduino IDE that you can download from the following URL: https://www.arduino.cc/en/main/software.

Note that we are going to use the Arduino IDE in all the projects of this book, so make sure to install the latest version.

Hardware configuration

We are now going to assemble the hardware for this project. As this is the first project of this book, it will be quite simple. However, there are quite a lot of components, so be sure to follow all the steps.

Here is a schematic to help you out during the process:

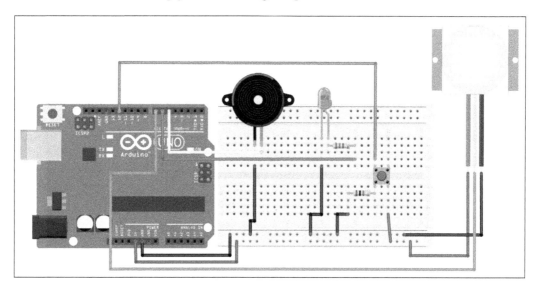

Let's start by putting all the components on the board. Place the buzzer, button, and LED on the board first, according to the schematics. Then, place the 330 Ohm resistor in series with the LED anode (the longest pin) and connect the 1k Ohm resistor to one pin of the push button.

This is how it should look at this stage:

Now we are going to connect each component to the Arduino board.

Let's start with the power supply. Connect the 5V pin of the Arduino board to one red power rail of the breadboard, and the GND pin of the Arduino board to one blue power rail of the breadboard.

Then, we are going to connect the buzzer. Connect one pin of the buzzer to pin number 5 of the Arduino board and the other pin to the blue power rail of the breadboard.

After that, let's connect the LED. Connect the free pin of the resistor to pin number 6 of the Arduino board and the free pin of the LED (the cathode) to the ground via the blue power rail.

Let's also connect the push button to our Arduino board. Refer to the schematic to be sure about the connections since it is a bit more complex. Basically, you need to connect the free pin of the resistor to the ground and connect the pin that is connected to the button to the 5V pin via the red power rail. Finally, connect the other side of the button to pin 12 of the Arduino board.

Finally, let's connect the PIR motion sensor to the Arduino board. Connect the VCC pin of the motion sensor to the red power rail and the GND pin to the blue power rail. Finally, connect the SIG pin (or OUT pin) to Arduino pin number 7.

The following is the final result:

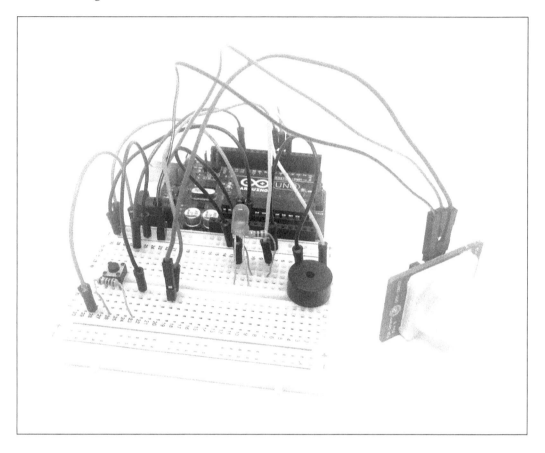

If your project looks similar to this picture, congratulations, you just assembled your first secret agent project! You can now go on to the next section.

Configuring the alarm system

Now that the hardware for our project is ready, we can write down the code for the project so that we have a usable alarm system. The goal is to make the buzzer produce a sound whenever motion is detected and also to make the LED flash. However, whenever the button is pressed, the alarm will be switched off.

Here is the complete code for this project:

```
// Code for the simple alarm system

// Pins
const int alarm_pin = 5;
const int led_pin = 6;
const int motion_pin = 7;
const int button_pin = 12;

// Alarm
boolean alarm_mode = false;

// Variables for the flashing LED
int ledState = LOW;
long previousMillis = 0;
long interval = 100;  // Interval at which to blink (milliseconds)

void setup()
{
  // Set pins to output
  pinMode(led_pin,OUTPUT);
  pinMode(alarm_pin,OUTPUT);

  // Set button pin to input
  pinMode(button_pin, INPUT);

  // Wait before starting the alarm
  delay(5000);
}

void loop()
{
  // Motion detected ?
  if (digitalRead(motion_pin)) {
    alarm_mode = true;
  }

  // If alarm mode is on, flash the LED and make the alarm ring
  if (alarm_mode){
    unsigned long currentMillis = millis();
    if(currentMillis - previousMillis > interval) {
      previousMillis = currentMillis;
      if (ledState == LOW)
```

```
        ledState = HIGH;
      else
        ledState = LOW;
    // Switch the LED
    digitalWrite(led_pin, ledState);
    }
    tone(alarm_pin,1000);
  }

  // If alarm is off
  if (alarm_mode == false) {

    // No tone & LED off
    noTone(alarm_pin);
    digitalWrite(led_pin, LOW);
  }

  // If button is pressed, set alarm off
  int button_state = digitalRead(button_pin);
  if (button_state) {alarm_mode = false;}
}
```

> **Downloading the example code**
>
> You can download the example code files from your account at
> http://www.packtpub.com for all the Packt Publishing books you
> have purchased. If you purchased this book elsewhere, you can visit
> http://www.packtpub.com/support and register to have the
> files e-mailed directly to you.

We are now going to see, in more detail, the different parts of the code. It starts by declaring which pins are connected to different elements of the project, such as the alarm buzzer:

```
const int alarm_pin = 5;
const int led_pin = 6;
const int motion_pin = 7;
const int button_pin = 12;
```

After that, in the `setup()` function of the sketch, we declare these pins as either inputs or outputs, as follows:

```
// Set pins to output
pinMode(led_pin,OUTPUT);
pinMode(alarm_pin,OUTPUT);

// Set button pin to input
pinMode(button_pin, INPUT);
```

Then, in the `loop()` function of the sketch, we check whether the alarm was switched on by checking the state of the motion sensor:

```
if (digitalRead(motion_pin)) {
  alarm_mode = true;
}
```

Note that if we detect some motion, we immediately set the `alarm_mode` variable to true. We will see how the code makes use of this variable right now.

Now, if the `alarm_mode` variable is true, we have to enable the alarm, make the buzzer emit a sound, and also flash the LED. This is done by the following code snippet:

```
if (alarm_mode){
    unsigned long currentMillis = millis();
    if(currentMillis - previousMillis > interval) {
      previousMillis = currentMillis;
      if (ledState == LOW)
        ledState = HIGH;
      else
        ledState = LOW;
    // Switch the LED
    digitalWrite(led_pin, ledState);
    }
    tone(alarm_pin,1000);
  }
```

Also, if `alarm_mode` is returning false, we need to deactivate the alarm immediately by stopping the sound from being emitted and shutting down the LED. This is done with the following code:

```
if (alarm_mode == false) {

    // No tone & LED off
    noTone(alarm_pin);
    digitalWrite(led_pin, LOW);
  }
```

Finally, we continuously read the state of the push button. If the button is pressed, we will immediately set the alarm off:

```
int button_state = digitalRead(button_pin);
if (button_state) {alarm_mode = false;}
```

Usually, we should take care of the bounce effect of the button in order to make sure that we don't have erratic readings when the button is pressed. However, here we only care about the button actually being pressed so we do not need to add an additional debouncing code for the button.

Note that you can find all the code for this project inside the GitHub repository of the book:

```
https://github.com/marcoschwartz/arduino-secret-agents
```

Now that we have written down the code for the project, it's time to get to the most exciting part of the chapter: testing the alarm system!

Testing the alarm system

We are now ready to test our simple alarm system. Just grab the code for this project (either from the preceding code or the GitHub repository of the book) and put it into your Arduino IDE.

In the IDE, choose the right board type (for example, Arduino Uno) and also the correct serial port.

You can now upload the code to the board. Once it is done, simply pass your hand in front of the PIR motion sensor; the alarm should go off immediately. Then, simply press the push button to stop it.

To illustrate the behavior of the alarm, I simply used a battery pack to make it work when it is not connected to my computer. The following is the result when the alarm goes off:

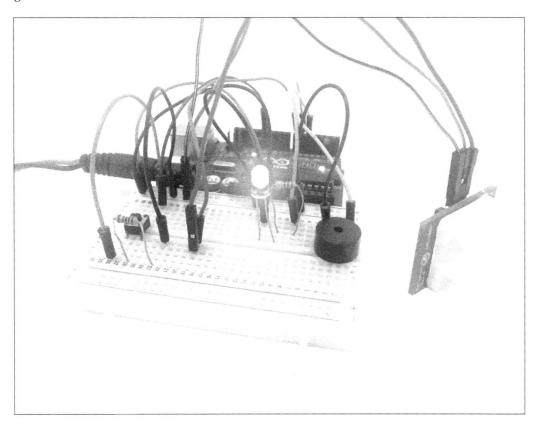

If this works as expected, congratulations, you just built your first secret agent project: a simple alarm system based on Arduino!

If it doesn't work well at this point, there are several things you can check. First, go through the hardware configuration part again to make sure that your project is correctly configured.

Also, you can verify that when you pass your hand in front of the PIR sensor, it goes red. If this is not the case, most probably your PIR motion sensor has a problem and must be replaced.

Summary

In this first chapter, we built a simple alarm based on Arduino with only a few components.

There are several ways to go further and improve this project. You can add more functions to the project just by adding more lines to the code. For example, you can add a timer so that the alarm only goes off after a given amount of time, or you can build a mode where a push of the button actually activates or deactivates the alarm mode.

In the next chapter, we are going to build another project that is very useful for secret agents: an audio recording device based on Arduino!

Creating a Spy Microphone

2

In this chapter, we are going to build a very useful device for any secret agent: a spy microphone. The project will be based on Arduino, with a simple amplified microphone and an SD card.

The following are the steps that we are going to take to build this project:

- We will see how to configure the project in order to make sure that it is recording for a given amount of time that can be configured by the user
- Then, the recorded audio file will be written on the SD card and be accessible from any computer
- Before doing that, we will test all the components of the project individually

Let's dive in!

Hardware and software requirements

Let's first see what the required components for this project are. As usual, we will use an Arduino Uno board as the 'brain' of the project.

Then, we will need a microphone. I used a simple SparkFun electret microphone, which has an amplifier onboard, as shown in the following image:

The most important thing here is that the microphone is amplified. For example, SparkFun is amplified 100 times, making it possible for the Arduino Uno to record usual sound levels (such as voices).

Then, you will need a microSD card with an adapter:

You will also need a way to record data on the SD card. There are many ways to do so with Arduino. The easiest, which is the solution that I chose here, is to use a shield. I had an Ethernet Shield available, which is great because it also has an onboard microSD card reader.

You can, of course, use any shield with a microSD card reader or even a microSD reader breakout board.

You will also need a breadboard and some jumper wires to make the required connections.

Finally, the following is the list of all the components that we will use in this project:

- Arduino Uno (`https://www.sparkfun.com/products/11021`)
- Arduino Ethernet Shield (`https://www.sparkfun.com/products/11166`)
- Electret Microphone (`https://www.sparkfun.com/products/9964`)
- microSD card (`https://www.sparkfun.com/products/11609`)
- Breadboard (`https://www.sparkfun.com/products/12002`)
- Jumper wires (`https://www.sparkfun.com/products/8431`)

On the software side, you will need a special version of the SD card library called `SdFat`. We can't use the usual Arduino SD library here as we will do some really fast write operations on the SD card, which can't be handled by the SD library that comes with the Arduino software. You can download this library from `https://github.com/greiman/SdFat`.

Using the SD card

The first thing that we are going to do in this project is to test whether we can actually access the SD card. This will ensure that we don't run into SD card-related problems later in the project.

This is a picture of the Ethernet Shield that I used, with the microSD card mounted on the right:

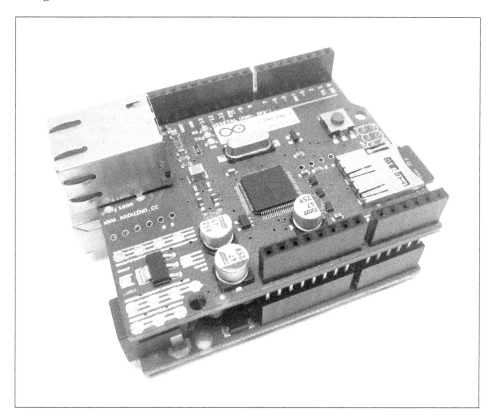

Let's now see the code that we will use to test the SD card's functionalities. The following is the complete code for this section:

```
// Include the SD library
#include <SPI.h>
#include <SD.h>

// Set up variables using the SD utility library functions:
Sd2Card card;
SdVolume volume;
SdFile root;

// change this to match your SD shield or module;
// Arduino Ethernet shield: pin 4
// Adafruit SD shields and modules: pin 10
// Sparkfun SD shield: pin 8
```

```
const int chipSelect = 4;

void setup()
{
  // Open serial communications and wait for port to open:
  Serial.begin(115200);
  while (!Serial) {
    ; // wait for serial port to connect. Needed for Leonardo only
  }


  Serial.print("\nInitializing SD card...");

  // we'll use the initialization code from the utility libraries
  // since we're just testing if the card is working!
  if (!card.init(SPI_HALF_SPEED, chipSelect)) {
    Serial.println("initialization failed. Things to check:");
    Serial.println("* is a card inserted?");
    Serial.println("* is your wiring correct?");
    Serial.println("* did you change the chipSelect pin to match your
shield or module?");
    return;
  } else {
    Serial.println("Wiring is correct and a card is present.");
  }

  // print the type of card
  Serial.print("\nCard type: ");
  switch (card.type()) {
    case SD_CARD_TYPE_SD1:
      Serial.println("SD1");
      break;
    case SD_CARD_TYPE_SD2:
      Serial.println("SD2");
      break;
    case SD_CARD_TYPE_SDHC:
      Serial.println("SDHC");
      break;
    default:
      Serial.println("Unknown");
  }

  // Now we will try to open the 'volume'/'partition' - it should be
FAT16 or FAT32
```

```
  if (!volume.init(card)) {
    Serial.println("Could not find FAT16/FAT32 partition.\nMake sure
you've formatted the card");
    return;
  }


  // print the type and size of the first FAT-type volume
  uint32_t volumesize;
  Serial.print("\nVolume type is FAT");
  Serial.println(volume.fatType(), DEC);
  Serial.println();

  volumesize = volume.blocksPerCluster();    // clusters are
collections of blocks
  volumesize *= volume.clusterCount();        // we'll have a lot of
clusters
  volumesize *= 512;                          // SD card blocks are
always 512 bytes
  Serial.print("Volume size (bytes): ");
  Serial.println(volumesize);
  Serial.print("Volume size (Kbytes): ");
  volumesize /= 1024;
  Serial.println(volumesize);
  Serial.print("Volume size (Mbytes): ");
  volumesize /= 1024;
  Serial.println(volumesize);


  Serial.println("\nFiles found on the card (name, date and size in
bytes): ");
  root.openRoot(volume);

  // list all files in the card with date and size
  root.ls(LS_R | LS_DATE | LS_SIZE);
}


void loop(void) {

}
```

This code tests a lot of things on the SD card, such as the file format and available space, and also lists all the files present on the SD card. However, what we are really interested in is to know whether the SD card can be read by the Arduino board. This is done with the following code snippet:

```
if (!card.init(SPI_HALF_SPEED, chipSelect)) {
    Serial.println("initialization failed. Things to check:");
    Serial.println("* is a card inserted?");
    Serial.println("* is your wiring correct?");
    Serial.println("* did you change the chipSelect pin to match your
shield or module?");
    return;
  } else {
    Serial.println("Wiring is correct and a card is present.");
  }
```

Now, let's test the code. You can simply copy this code and paste it into the Arduino IDE.

Then, upload it to the Arduino Uno board and open the serial monitor. Make sure that the serial speed is set to 115200 bps. This is what you will see:

```
                            /dev/cu.usbmodem1a12121

                                                              Send

Initializing SD card...Wiring is correct and a card is present.

Card type: SDHC

Volume type is FAT32

Volume size (bytes): 3665821696
Volume size (Kbytes): 3579904
Volume size (Mbytes): 3496

Files found on the card (name, date and size in bytes):
~1.TRA          2015-07-27 10:26:58 4096
REC00000.WAV    2000-01-01 01:00:00 211500
TRASHE~1/       2015-07-27 10:26:58

✓ Autoscroll                    Carriage return  ⁝    115200 baud  ⁝
```

If you can see this, congratulations, your SD card and card reader are correctly configured and ready to host some spy audio recordings!

Testing the microphone

We are now going to make sure that the microphone is working correctly and especially check whether it can record voice levels, for example. I had a problem when I was testing the prototype of this project with a microphone that wasn't amplified; I just couldn't hear anything on the recording.

The first step is to plug the microphone into the Arduino board. There are 3 pins to connect the microphone: VCC, GND, and AUD. Connect VCC to the Arduino 5V pin, GND to the Arduino GND pin, and AUD to the Arduino analog pin A5.

The following is a schematic to help you out:

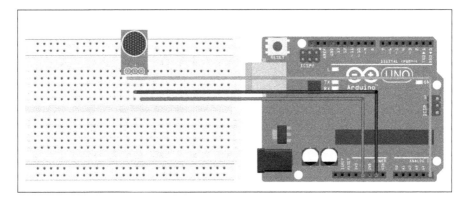

Here is an image of the final result:

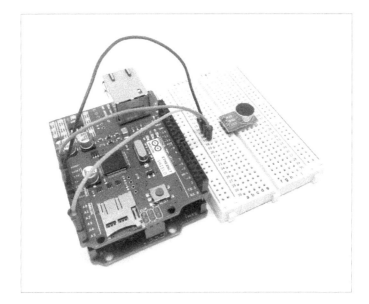

We are now going to use a very simple sketch to read out the signal from the microphone and print it on the serial monitor:

```
// Microphone test

void setup() {

  // Start Serial
  Serial.begin(115200);
}

void loop() {

  // Read the input on analog pin 5:
  int sensorValue = analogRead(A5);

  // Print out the value you read:
  Serial.println(sensorValue);
  delay(1);          // delay in between reads for stability
}
```

This sketch, basically, continuously reads the data from the A5 pin, where the microphone is connected, and prints it on the serial monitor.

Now, copy and paste this sketch in the Arduino IDE and upload it to the board. Also, open the serial monitor.

The following is the result on the serial monitor:

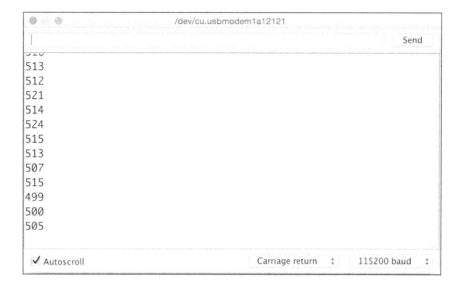

While looking at the serial monitor, speak around the microphone. You should immediately see some variations in the signal that is read by the board. This means that your voice is being recorded by the microphone and the amplification is sufficient for the microphone to record a normal voice level.

Building the spy microphone

In this section, we are going to put everything together and actually build our spy microphone.

The hardware for the project is nearly ready if you followed the previous section. You just need to plug the SD card into the reader again.

I also added an LED on pin 7, just to know when the recording is on. If you want to do the same, you just need an LED and a 330 Ohm resistor. Of course, remove this LED when you actually want to use it as a spy microphone or your project might get noticed.

The schematic to help you out is as follows:

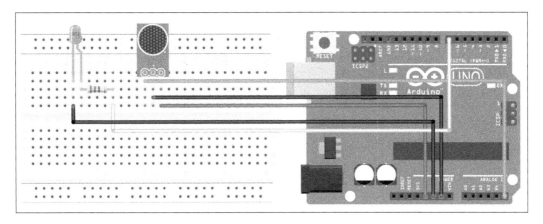

The following is the image of the completely assembled project:

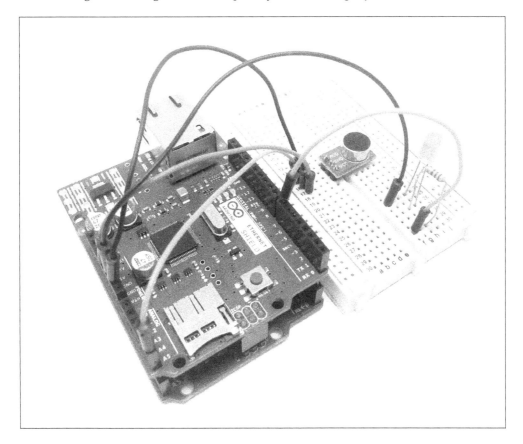

We are now going to see the details of the code for the project. Basically, we want the device to record audio from the microphone for a given amount of time and then stop the recording.

As the code is long and complex, we are only going to see the most important parts here.

The first step is to include the `SdFat` library:

```
#include <SdFat.h>
```

Then, we will create the instances that are necessary in order to write on the SD card:

```
SdFat sd;
SdFile rec;
```

After that, we will define on which pin we want the optional recording LED to be:

```
const int ledStart = 7;
```

We will also define a variable for the internal counter of the project:

```
byte recordingEnded = false;
unsigned int counter;
unsigned int initial_count;
unsigned int maxCount = 10 * 1000; // 10 Seconds
```

Note that here you will have to modify the `maxCount` variable according to the time for which you want the device to record. Here, I just used 10 seconds by default as a test.

Then, we will initialize the ADC (Analog-Digital Converter):

```
Setup_timer2();
Setup_ADC();
```

After that, we will initialize the SD card and also flash the LED if this is successful:

```
if (sd.begin(chipSelect, SPI_FULL_SPEED)) {
  for (int dloop = 0; dloop < 4; dloop++) {
    digitalWrite(ledStart, !digitalRead(ledStart));
    delay(100);
  }
}
```

Then, we will actually start the recording with the following function:

```
StartRec();
```

We will also initialize the counter with the following line of code:

```
initial_count = millis();
```

Then, in the `loop()` function of the sketch, we will update the counter, which helps to keep track of the elapsed time:

```
counter = millis() - initial_count;
```

Still in the `loop()` function, we will actually stop the recording if we reach the maximum amount of time that we defined previously:

```
if (counter > maxCount && !recordingEnded) {
  recordingEnded = true;
  StopRec();
}
```

Note that the whole code for this section can be found in the GitHub repository of the book at `https://github.com/marcoschwartz/arduino-secret-agents`.

Recording on the SD card

In the last section of the chapter, we are actually going to test the project and record some audio.

First, copy all the code and paste it into the Arduino IDE. Compile it and upload it to the Arduino board. Note that, as soon as you do that, the project will start recording the audio. If you have connected the optional LED on pin 7, the LED should also be on during the recording phase.

You can now talk a bit or play your favorite song just to make sure that actual audio is being recorded by the microphone.

Then, after the amount of time defined in the code, stop the project by disconnecting the power. Then, remove the SD card and insert it into your computer.

On your computer, navigate to the SD card and you will see that one file was recorded:

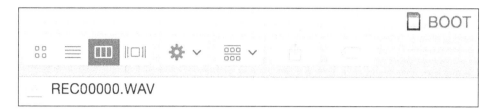

You can now simply open this file with your favorite audio player and listen to what was just recorded.

I, for example, opened it with the free audio editing software, Audacity, to see how the waveform looked like:

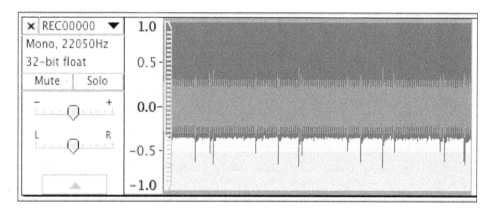

Congratulations, you just built your own spy microphone! You can now play with the different settings, for example, using a longer recording time.

Summary

In this chapter, you learned how to build a spy microphone based on a simple amplified microphone, Arduino, and a microSD card. The spy microphone can be set up to record audio continuously for a given amount of time.

Of course, there are many things that you can now do in order to improve this project. You can, for example, use a battery to power the project to make it autonomous. Then, you just need to place it in a room in which you want to record a conversation and just come back later to retrieve the SD card with the recording.

You can also connect a motion sensor to the project and use that to automatically start a recording when motion is detected in a room in order to intercept conversations with total discretion.

Another interesting project would be to use the audio levels that are sensed by the project to actually start or stop the recording. For example, you could rewrite the software of the project to continuously monitor the microphone and automatically start a recording when a given threshold is crossed, indicating that somebody is speaking. Then, the program could automatically stop the recording once the audio levels are low again.

In the next chapter of the book, we are going to make another useful project for any spy: an EMF detector to detect the presence of a recording device in a room, for example.

3
Building an EMF Bug Detector

In this chapter, we are going to build a very useful tool that every secret agent should have: a bug detector. We will build a simple device that will allow you to detect whether there are any bugs nearby, such as a recording device or a wireless hidden camera.

The following are the topics that we will cover in this chapter:

- We will build a project with a simple wire antenna and the project will display the EMF readings on LCD screen.
- We will also add a simple LED to indicate when EMF activity goes above a certain threshold.

Let's dive in!

Hardware and Software requirements

First, let's see what the required components for this project are.

You will, of course, need the usual Arduino Uno, that will act as the brain of the project and process all the information.

You will also need a simple wire, preferably long (such as 10 cm to 20 cm) to act as an antenna. You will need a 1M Ohm resistor along with this wire.

To display the data, we will use a simple I2C LCD screen. I used a 4 x 20 I2C screen from DFRobot:

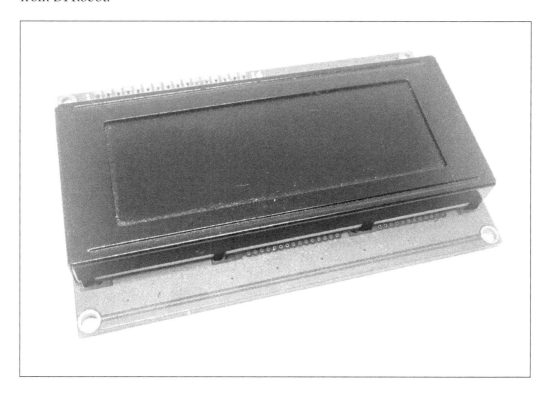

You can, of course, use the LCD screen of your choice for this project, you will just need to use the right LCD library.

I also integrated a simple red LED along with a 330 Ohm resistor to display when EMF goes above a given level.

Finally, you will also need a breadboard and jumper wires to make the different connections.

Here is the list of all the components that we will use for this project:

- Arduino Uno (https://www.sparkfun.com/products/11021)
- I2C LCD screen (http://www.robotshop.com/en/dfrobot-i2c-twi-lcd-module.html)
- Red LED (https://www.sparkfun.com/products/9590)
- 330 Ohm resistor (https://www.sparkfun.com/products/8377)
- Long wire (at least 10 cm)

- 1M Ohm resistor (https://www.sparkfun.com/products/11853)
- Breadboard (https://www.sparkfun.com/products/12002)
- Jumper wires (https://www.sparkfun.com/products/8431)

On the software side, you will need the library for the LCD screen. As we will use an I2C LCD screen for this project, I recommend the following library that you can download from http://hmario.home.xs4all.nl/arduino/LiquidCrystal_I2C/.

Once the library is correctly installed, you can move to the next step.

Hardware configuration

We are now going to configure the hardware part of the project. As we have a relatively small number of simple components, the configuration of this project will be really easy and straightforward. This is a schematic to help you out:

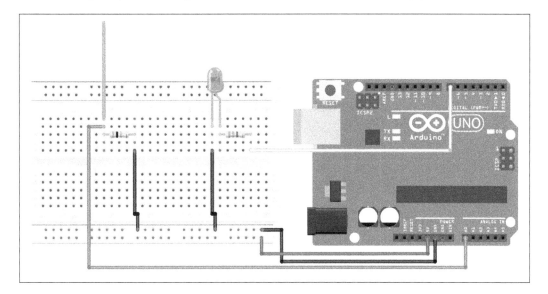

As you can see, the LCD screen is not present on this schematic. We'll first see how to connect all the other components and then see how to connect the LCD screen at the end of this section.

First, connect the power to the breadboard: connect GND to the blue power rail of the breadboard and the +5V pin to the red power rail.

Then, we are going to connect the antenna: first, place it on the breadboard in series with the 1M Ohm resistor. Then, connect the other end of the resistor to the ground. Finally, connect the antenna to the A0 analog pin.

For the LED, simply place it in series with the resistor, as seen on the schematic. Ensure that you connect the resistor to the anode of the LED, which is the longest pin of the LED. Finally, connect the other end of the resistor to the digital pin 7 and the other end of the LED to ground.

For the LCD screen, the connections are really easy, thanks to the I2C interface. First connect the power: VCC goes to the red power rail and GND to the blue power rail. For the data pin, connect SDA and SCL to their respective pins on the Arduino board, which are next to the digital pin 13.

The following image is the final result:

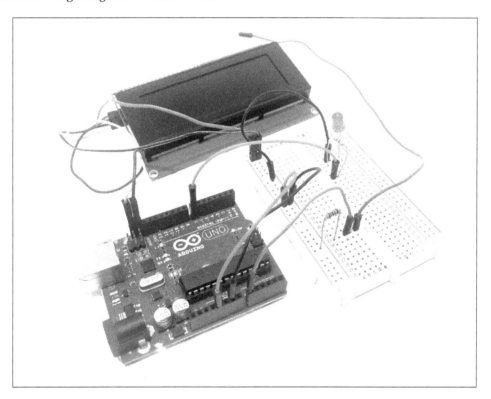

Congratulations! We are now going to see how to test the LCD screen.

Testing the LCD screen

Before we build our EMF bug detector, we want to make sure that the LCD screen is working correctly. Therefore, we are going to test it by printing a very simple message on it.

The following is a complete sketch to do this:

```
// Required libraries
#include <Wire.h>
#include <LiquidCrystal_I2C.h>

// Create LCD instance
LiquidCrystal_I2C lcd(0x27,20,4);

void setup()
{
  // Initialise LCD
  lcd.init();

  // Print a message to the LCD
  lcd.backlight();
  lcd.setCursor(0,0);
  lcd.print("Hello Secret Agent!");
}


void loop()
{
}
```

As you can see, the sketch is pretty straightforward. You can now plug the Arduino project into your computer using a USB cable and then copy and paste the sketch into your Arduino IDE.

Then, upload the sketch to your board. This is what you should see:

If you can see this message, congratulations, you are ready to move to the next step!

Building the EMF bug detector

We are now going to dive into the core of this project and configure the project so that it can detect EMF activity around the antenna.

The following is the complete code for this project:

```
// Required libraries
#include <Wire.h>
#include <LiquidCrystal_I2C.h>

// Number of readings
#define NUMREADINGS 15

// Parameters for the EMF detector
int senseLimit = 15;
int probePin = 0;
int ledPin = 7;
int val = 0;
int threshold = 200;

// Averaging the measurements
```

```
int readings[NUMREADINGS];
int index = 0;
int total = 0
int average = 0;

// Time between readings
int updateTime = 40;

// Create LCD instance
LiquidCrystal_I2C lcd(0x27,20,4);

void setup()
{
  // Initialise LCD
  lcd.init();

  // Set LED as output
  pinMode(ledPin, OUTPUT);

  // Print a welcome message to the LCD
  lcd.backlight();
  lcd.setCursor(0,0);
  lcd.print("EMF Detector Started");
  delay(1000);
  lcd.clear();
}


void loop()
{
  // Read from the probe
  val = analogRead(probePin);
  Serial.println(val);

  // Check reading
  if(val >= 1){

    // Constrain and map with sense limit value
    val = constrain(val, 1, senseLimit);
    val = map(val, 1, senseLimit, 1, 1023);

    // Averaging the reading
    total -= readings[index];
    readings[index] = val;
```

```
      total += readings[index];
      index = (index + 1);

      if (index >= NUMREADINGS)
        index = 0;

      average = total / NUMREADINGS;

      // Print on LCD screen
      lcd.setCursor(0,1);
      lcd.print("   ");

      lcd.setCursor(0,0);
      lcd.print("EMF level: ");
      lcd.setCursor(0,1);
      lcd.print(average);

      // Light up LED if EMF activity detected
      if (average > threshold) {
        digitalWrite(ledPin, HIGH);
      }
      else {
        digitalWrite(ledPin, LOW);
      }

      // Wait until next reading
      delay(updateTime);
    }
  }
```

Now let's see the details of this code. It starts by including the required libraries:

```
#include <Wire.h>
#include <LiquidCrystal_I2C.h>
```

Then, we define the number of readings that we want to do at every iteration in order to make sure that we get the average EMF reading:

```
#define NUMREADINGS 15
```

We will then define some parameters for the project, such as the different pins on which the components are connected:

```
int senseLimit = 15;
int probePin = 0;
int ledPin = 7;
int val = 0;
int threshold = 200;
```

After that, we will define some variables that will be used to take the average of the readings:

```
int readings[NUMREADINGS];
int index = 0;
int total = 0;
int average = 0;
```

We will also set an update time that we will leave between each EMF reading. By default, we set it to 40 milliseconds:

```
int updateTime = 40;
```

We will also create an instance of the LCD screen:

```
LiquidCrystal_I2C lcd(0x27,20,4);
```

After that, in the setup() function of the sketch, we will initialize the LCD screen:

```
// Initialise LCD
  lcd.init();

  // Set LED as output
  pinMode(ledPin, OUTPUT);

  // Print a welcome message to the LCD
  lcd.backlight();
  lcd.setCursor(0,0);
  lcd.print("EMF Detector Started");
  delay(1000);
  lcd.clear();
```

Then, in the loop() function of the sketch, we will get the value of the analog pin on which the antenna is connected:

```
val = analogRead(probePin);
```

After that, we will constrain the reading to the limit value that was defined previously, and then map it again between 1 and 1023:

```
val = constrain(val, 1, senseLimit);
val = map(val, 1, senseLimit, 1, 1023);
```

Then, we will take an average of the readings:

```
total -= readings[index];
readings[index] = val;
total += readings[index];
index = (index + 1);

if (index >= NUMREADINGS)
  index = 0;
average = total / NUMREADINGS;
```

Finally, we will display the average on the LCD screen and also light up the LED if the average reading is higher than the threshold:

```
lcd.setCursor(0,1);
lcd.print("    ");

lcd.setCursor(0,0);
lcd.print("EMF level: ");
lcd.setCursor(0,1);
lcd.print(average);

// Light up LED if EMF activity detected
if (average > threshold) {
  digitalWrite(ledPin, HIGH);
}
else {
  digitalWrite(ledPin, LOW);
}
```

We will also wait for a given amount of time between each reading:

```
delay(updateTime);
```

Note that you can find the complete code in the GitHub repository of the project at `https://github.com/marcoschwartz/arduino-secret-agents`.

It's now time to test the project. You can copy the complete code and paste it into your Arduino IDE, or just get it from GitHub.

Then, upload the code to your project. After a while, you should see that the LCD screen is being initialized and is blank. This is because it is waiting to detect a significant EMF activity.

I, for example, approached my phone with the antenna and got the following result:

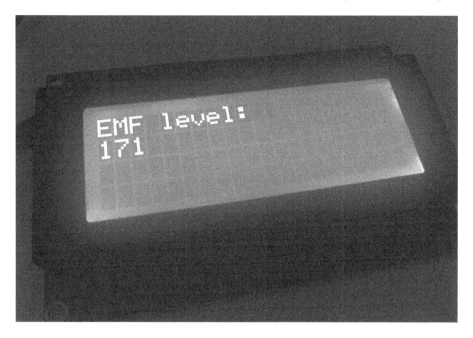

If you can see something on your screen, congratulations, you just built an EMF bug detector! Now, I invite you to try with other devices, such as a surveillance camera, any Wi-Fi device, a radio controller, and so on.

Summary

In this chapter, we built a simple EMF bug detector that a secret agent can use to see whether there are any bugs present in a room, such as an audio recorder or a spy camera.

There are, of course, several ways to improve this project. You can, for example, connect a battery to the project and integrate it in a small case in order to build a simple and portable EMF bug detector.

In the next chapter, we are going to build another useful device for a secret agent. We will see how to use a fingerprint scanner to secure access to important information.

Access Control with a Fingerprint Sensor

In this chapter, we are going to build a more complex secret agent project: an access control system using a fingerprint sensor. We will connect the fingerprint sensor to Arduino along with a relay and an LCD screen.

Based on this hardware, we will build several cool projects. The following are the steps that we will take for this project:

- First, we are going to record your fingerprint in the sensor so that you can get access
- Then, we will use the fingerprint sensor to open or close the relay
- Finally, we will create a system with the LCD screen to grant access to a secret piece of data stored in Arduino

Let's dive in!

Hardware and software requirements

First, let's see with fingerprint sensor: software requisites "what with fingerprint sensor:hardware requisites" are the required components for this project.

As usual, we will use an Arduino Uno board as the *brain* of the project.

The most important part of this project is the fingerprint sensor. The following is an image of the sensor that I used:

You with fingerprint sensor:software requisites "need to get with fingerprint sensor:hardware requisites" the exact same model (from Adafruit). Otherwise, the code for this project won't work.

You will also need an LCD screen for the last part of this project. I used an I2C LCD screen from DFRobot that we already used earlier in the book.

I also used a Pololu 5V relay module, which is really convenient to connect to Arduino. A relay will basically allow us to control a wide range of devices, for example, from a simple LED to electrical appliances.

Finally, here is the list of all the components that we will use in this project:

- Arduino Uno (https://www.sparkfun.com/products/11021)
- Adafruit Fingerprint Sensor (https://www.adafruit.com/products/751)
- DFRobot 4x20 LCD screen (http://www.robotshop.com/en/dfrobot-i2c-twi-lcd-module.html)

- Pololu relay module (https://www.pololu.com/product/2480)
- Breadboard (https://www.sparkfun.com/products/12002)
- Jumper wires (https://www.sparkfun.com/products/8431)

On the software side, you will need two Arduino libraries: the `LiquidCrystal_I2C` library with fingerprint sensor:software requisites "for the LCD with fingerprint sensor:hardware requisites" screen and the `Adafruit Fingerprint Sensor` library. You can get them both using the Arduino library manager.

You can also visit the GitHub repository of the Fingerprint Sensor library, to learn more about the different functions, available at https://github.com/adafruit/Adafruit-Fingerprint-Sensor-Library.

Hardware configuration

We are first going to see how to assemble the different parts of this project.

Let's start by connecting the power supply. Connect the 5V pin from the Arduino board to the red power rail and the GND from Arduino to the blue power rail on the breadboard.

Now, let's connect the fingerprint sensor. First, connect the power by connecting the cables to their respective color on the breadboard. Then, connect the white wire from the sensor to Arduino pin 3 and the green wire to pin number 2.

After that, we are going to connect the relay module. Connect the VCC pin to the red power rail, GND pin to the blue power rail, and the EN pin to Arduino pin 7.

Finally, let's connect the LCD screen. First, connect the power: VCC to the red power rail and GND pin to the blue power rail. After that, connect the I2C pins (SDA and SCL) to the Arduino board. The I2C pins are next to pin 13 on the Arduino Uno board.

The following is the final result:

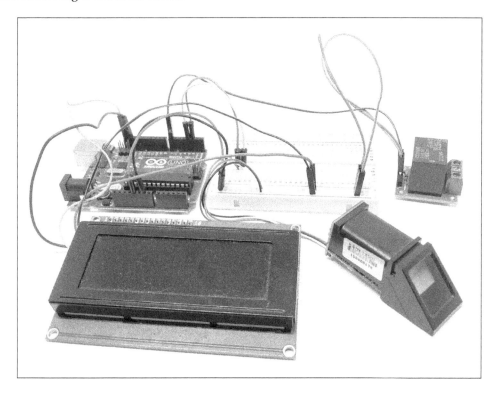

If your project is similar to the preceding image, congratulations, you can now proceed to the next section of this chapter and build exciting applications based on the fingerprint sensor!

Enrolling your fingerprint

The first thing that we have to do is enroll at least one fingerprint so that it can be later recognized by the sensor. We will do that in this section. Here is most of the code for this section:

```
// Libraries
#include <Adafruit_Fingerprint.h>
#include <SoftwareSerial.h>

// Fingerprint enroll function
uint8_t getFingerprintEnroll(uint8_t id);

// Software Serial instance
```

```
SoftwareSerial mySerial(2, 3);

// Fingerprint sensor instance
Adafruit_Fingerprint finger = Adafruit_Fingerprint(andmySerial);

void setup()
{
  // Start serial
  Serial.begin(9600);
  Serial.println("fingertest");

  // Set the data rate for the sensor serial port
  finger.begin(57600);

  // Verify that sensor is present
  if (finger.verifyPassword()) {
    Serial.println("Found fingerprint sensor!");
  } else {
    Serial.println("Did not find fingerprint sensor :(");
    while (1);
  }
}

void loop()
{
  // Wait for fingerprint ID
  Serial.println("Type in the ID # you want to save this finger
as...");
  uint8_t id = 0;
  while (true) {
    while (! Serial.available());
    char c = Serial.read();
    if (! isdigit(c)) break;
    id *= 10;
    id += c - '0';
  }
  Serial.print("Enrolling ID #");
  Serial.println(id);

  while (!  getFingerprintEnroll(id) );
}
```

You will note that I didn't include all the details of fingerprint sensor functions as they are too long to be displayed here. You can, of course, find the whole code in the GitHub repository of this book.

Now, let's see the details of the code. It starts by including the required libraries:

```
#include <Adafruit_Fingerprint.h>
#include <SoftwareSerial.h>
```

Then, we declare the function that we will use in order to enroll our fingerprint:

```
uint8_t getFingerprintEnroll(uint8_t id);
```

After that, we create the software serial instance that we will use to communicate with the server:

```
SoftwareSerial mySerial(2, 3);
```

We will also create the fingerprint sensor instance:

```
Adafruit_Fingerprint finger = Adafruit_Fingerprint(andmySerial);
```

Now, in the `setup()` function of the sketch, we will initialize serial communications:

```
Serial.begin(9600);
Serial.println("fingertest");
```

Then, we will initialize the communication with the sensor:

```
finger.begin(57600);
```

We will also check whether the sensor is present:

```
if (finger.verifyPassword()) {
  Serial.println("Found fingerprint sensor!");
} else {
  Serial.println("Did not find fingerprint sensor :(");
  while (1);
}
```

In the `loop()` function of the code, we will first wait for an input from the user, which is the ID of the fingerprint that we want to store. Then, we go through the process of storing the fingerprint. This is all done by the following code snippet:

```
// Wait for fingerprint ID
Serial.println("Type in the ID # you want to save this finger as...");
  uint8_t id = 0;
  while (true) {
    while (! Serial.available());
```

```
    char c = Serial.read();
    if (! isdigit(c)) break;
    id *= 10;
    id += c - '0';
  }
  Serial.print("Enrolling ID #");
  Serial.println(id);

  while (!  getFingerprintEnroll(id) );
```

It's now the time to test the enrollment process. First, get the complete code, for example, from the GitHub repository of the book, which is available at `https://github.com/marcoschwartz/arduino-secret-agents`.

Then, copy the code in the Arduino IDE. After that, upload the code to the Arduino board and open the serial monitor with a speed of 9600 baud. The following screenshot is what you should see:

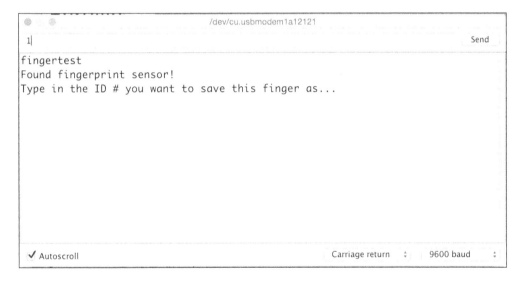

On being prompted, enter the ID of the fingerprint that you want to store and press enter. The sketch will now ask you to put your finger on the sensor. Do so and, after a while, you should see that the image was taken and you can now remove your finger:

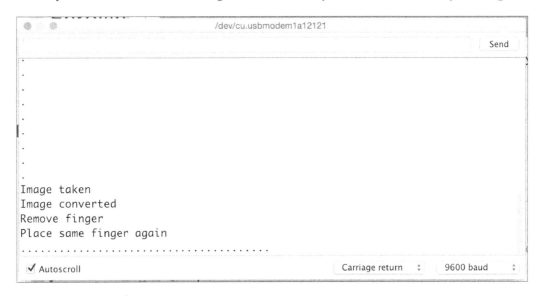

Then, as asked by the sketch, put your finger on the sensor once more. The sketch will then confirm that the fingerprint has been stored:

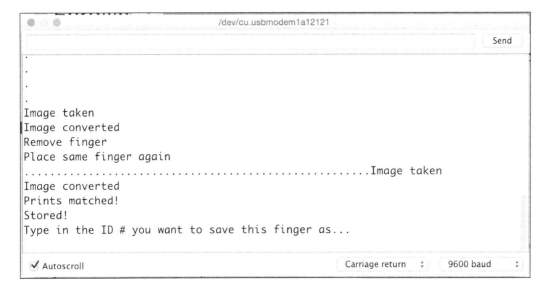

If you can see this message, it means that your fingerprint is now stored in the sensor and you can move to the next section.

Controlling access to the relay

Now that your fingerprint is stored in the sensor, you can build your first application using the hardware that we previously built. We are going to open or close the relay every time the sensor recognizes our fingerprint. The following is the complete code for this part, excluding the details about recognition functions:

```
// Libraries
#include <Adafruit_Fingerprint.h>
#include <SoftwareSerial.h>

// Function to get fingerprint
int getFingerprintIDez();

// Init Software Serial
SoftwareSerial mySerial(2, 3);

// Fingerprint sensor instance
Adafruit_Fingerprint finger = Adafruit_Fingerprint(andmySerial);

// Relay parameters
int relayPin = 7;
bool relayState = false;

// Your stored finger ID
int fingerID = 0;

// Counters
int activationCounter = 0;
int lastActivation = 0;

void setup()
{

  // Start Serial
  Serial.begin(9600);

  // Set the data rate for the sensor serial port
  finger.begin(57600);

  // Check if sensor is present
```

```
    if (finger.verifyPassword()) {
      Serial.println("Found fingerprint sensor!");
    } else {
      Serial.println("Did not find fingerprint sensor :(");
      while (1);
    }
    Serial.println("Waiting for valid finger...");

    // Set relay as output
    pinMode(relayPin, OUTPUT);
  }

void loop()
{
  // Get fingerprint # ID
  int fingerprintID = getFingerprintIDez();

  // Activation ?
  if ( (activationCounter - lastActivation) > 2000) {

    if (fingerprintID == fingerID && relayState == false) {
      relayState = true;
      digitalWrite(relayPin, HIGH);
      lastActivation = millis();
    }
    else if (fingerprintID == fingerID && relayState == true) {
      relayState = false;
      digitalWrite(relayPin, LOW);
      lastActivation = millis();
    }

  }
  activationCounter = millis();
  delay(50);
}
```

As you can see, many elements are common with the sketch that we saw in the previous section. We are only going to see those elements which are important and are added to this new sketch.

We have to define on which pin the relay is connected and also state that the relay is off by default:

```
int relayPin = 7;
bool relayState = false;
```

Then, we will define the ID under which we stored the fingerprint in the previous section. I used 0 here as I stored my fingerprint with the ID number 0:

```
int fingerID = 0;
```

Also, we don't want the relay to continuously switch state when we have our finger on the sensor. Therefore, we need the following two variables to count 2 seconds before the state of the relay can be changed again:

```
int activationCounter = 0;
int lastActivation = 0;
```

We will then set the relay pin as an output:

```
pinMode(relayPin, OUTPUT);
```

Then, in the loop() function of the sketch, we check whether the sensor is reading any fingerprint ID that is already stored in the sensor:

```
int fingerprintID = getFingerprintIDez();
```

The following is the check for the activation period:

```
if ( (activationCounter - lastActivation) > 2000) {
```

We will then check whether the ID corresponds to the ID that we defined earlier and also check the state of the relay. If the ID corresponds to the ID that we entered in the sketch, we switch the state of the relay:

```
if (fingerprintID == fingerID && relayState == false) {
    relayState = true;
  digitalWrite(relayPin, HIGH);
  lastActivation = millis();
    }
  else if (fingerprintID == fingerID && relayState == true) {
  relayState = false;
  digitalWrite(relayPin, LOW);
  lastActivation = millis();
}
```

Finally, we will refresh the activation counter and wait 50 milliseconds until the next read:

```
activationCounter = millis();
delay(50);
```

It's now time to test the sketch. Get all the code, for example, from the GitHub repository of the book and then upload it to the board. Make sure that you change the ID in the sketch, corresponding to the fingerprint that you stored earlier.

Then, open the serial monitor with the speed of 9600 baud. Place the finger that you recorded previously on the sensor. You should immediately see the following in the serial monitor:

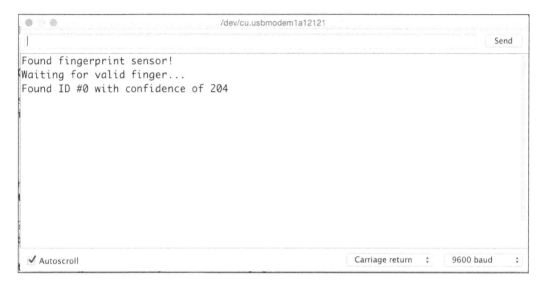

You should also hear the relay click, meaning that it just changed its state. You can now do the same operation after some seconds; you should hear the relay switch back to its initial state.

You can try with another finger or ask someone else to try the project; nothing will happen at all.

Accessing secret data

In the last section of this chapter, we are going to use all the hardware that we connected to the project for another cool application: accessing a secret piece of data with your fingerprint.

It can, for example, be a secret code that you only want to be accessible with your own fingerprint. We will use the LCD screen for this, removing the need to have the serial monitor opened.

The following is the complete code for this part, excluding the functions to read the fingerprint sensor data:

```
// Libraries
#include <Adafruit_Fingerprint.h>
#include <SoftwareSerial.h>
#include <LiquidCrystal_I2C.h>
#include <Wire.h>

// LCD Screen instance
LiquidCrystal_I2C lcd(0x27,20,4);

// Function to get fingerprint
int getFingerprintIDez();

// Init Software Serial
SoftwareSerial mySerial(2, 3);

// Fingerprint sensor instance
Adafruit_Fingerprint finger = Adafruit_Fingerprint(&mySerial);

// Relay parameters
int relayPin = 7;
bool relayState = false;

// Your stored finger ID
int fingerID = 0;

// Counters
int activationCounter = 0;
int lastActivation = 0;

// Secret data
String secretData = "u3fks43";

void setup()
{

  // Start Serial
  Serial.begin(9600);

  // Set the data rate for the sensor serial port
```

```
  finger.begin(57600);

  // Check if sensor is present
  if (finger.verifyPassword()) {
    Serial.println("Found fingerprint sensor!");
  } else {
    Serial.println("Did not find fingerprint sensor :(");
    while (1);
  }
  Serial.println("Waiting for valid finger...");

  // Set relay as output
  pinMode(relayPin, OUTPUT);

  // Init display
  lcd.init();
  lcd.backlight();
  lcd.clear();
  lcd.setCursor(0,0);
  lcd.print("Scan your finger");
}

void loop()
{
  // Get fingerprint # ID
  int fingerprintID = getFingerprintIDez();

  // Activation ?
  if ( (activationCounter - lastActivation) > 2000) {

    if (fingerprintID == fingerID) {

      lcd.clear();
      lcd.setCursor(0,0);
      lcd.print("Access granted!");
      lcd.setCursor(0,1);
      lcd.print("Your secret data is:");
      lcd.setCursor(0,2);
      lcd.print(secretData);

      if (relayState == false) {
        relayState = true;
        digitalWrite(relayPin, HIGH);
      }
```

```
    else if (relayState == true) {
      relayState = false;
      digitalWrite(relayPin, LOW);
    }
    lastActivation = millis();
  }
}
activationCounter = millis();
delay(50);
}
```

As you can see, this code has some common parts with the code that we saw in the earlier section. Therefore, we are only going to talk about the most important elements here.

It starts by including the required libraries:

```
#include <Adafruit_Fingerprint.h>
#include <SoftwareSerial.h>
#include <LiquidCrystal_I2C.h>
#include <Wire.h>
```

Then, we need to create an instance of the LiquidCrystal_I2C LCD screen:

```
LiquidCrystal_I2C lcd(0x27,20,4);
```

We will also define a string containing our secret data:

```
String secretData = "u3fks43";
```

Of course, feel free to put anything here. Then, in the setup() function, we will initialize the LCD screen:

```
lcd.init();
lcd.backlight();
lcd.clear();
lcd.setCursor(0,0);
lcd.print("Scan your finger");
```

Then, in the loop() function of the sketch, if we are outside of the activation period of 2 seconds, we will compare the fingerprint that was measured with the ID that we set in the sketch. If this is correct, we will print a message on the LCD saying that the access was granted along with the secret data:

```
if (fingerprintID == fingerID) {

    lcd.clear();
    lcd.setCursor(0,0);
```

```
lcd.print("Access granted!");
lcd.setCursor(0,1);
lcd.print("Your secret data is:");
lcd.setCursor(0,2);
lcd.print(secretData);
```

Now, it's time to test the sketch. Get all the code, for example, from the GitHub repository of the project and upload it to the board.

Then, place your finger that you previously recorded on the sensor. This is what you should see:

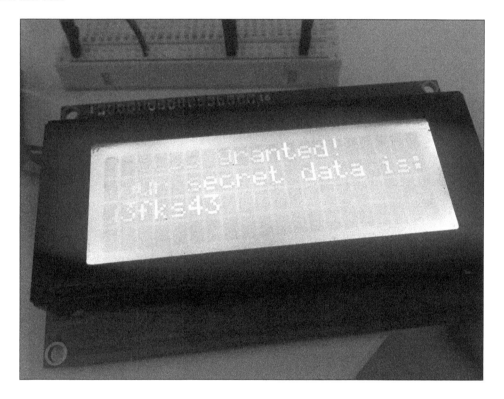

If you can see the preceding result, congratulations, you just built your own fingerprint access control system!

Summary

In this project, we built an access control system using a fingerprint sensor with Arduino. We also built several cool applications based on it, including a way to get access to a secret piece of data using your fingerprint.

There are, of course, many things you can do in order to improve this project. For example, you can code a way for the secret message to disappear after a given amount of time; avoiding someone else getting access to it after you used the project. You could also connect the board to the web (for example, with an Ethernet shield) and use the fingerprint sensor to send a given tweet when you put your finger on it.

In the next chapter, we are going to use another piece of hardware to build an access control system: a GSM shield that will allow us to open a door just by sending an SMS!

5
Opening a Lock with an SMS

In this chapter, we are going to build another great application for secret agents: opening a door lock simply by sending an SMS! You'll just need to send a message to a given number and then you'll be able to control a door lock or any other kind of on/off digital device, such as an alarm.

For that, we are going to take a number of steps, as follows:

- First, we are going to use Arduino and the FONA shield from Adafruit in order to be able to receive and process text messages
- Then, we'll connect this to a relay and LED to see whether we can actually control a device by sending an SMS
- Finally, we'll see how to connect an actual electronic lock to the system

Let's start!

Hardware and software requirements

First, let's see what the required components for this project are. The most important parts are related to GSM functionalities, which is the central piece in our project. We'll need an antenna in order to be able to connect to the local GSM network. For this project, a flat uFL antenna is used:

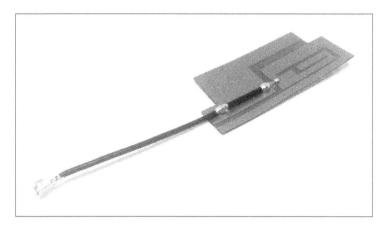

Then, you'll need a way to actually use a SIM card, connect to the GSM network, and process the information with Arduino. There are many boards that can do this; however, I recommend the Adafruit FONA shield, which is very convenient to configure and use with Arduino. The following is the image of the Adafruit FONA shield along with the flat GSM antenna.

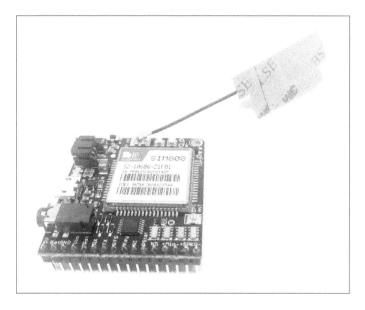

Then, you will need a battery to power the FONA shield, as the Arduino Uno board doesn't allow to power the chip that is at the core of the FONA shield (it can use up to 2A at a time!). For this, I used a 3.7 LiPo battery along with a micro USB battery charger:

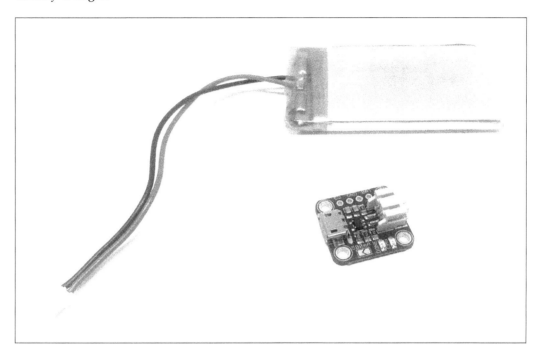

A very important part of the project is the SIM card that you need to place in the FONA shield. You need a normal SIM card (not micro or nano), which is activated, not locked by a PIN, and is able to receive text messages. You can get one at any of your local mobile network operators.

Then, to test the functionalities of the project, we'll also use a simple LED (along with a 330 Ohm resistor) and a 5V relay. This will mimic the behavior of the real electronic lock.

Finally, you'll need an electronic lock. This part is optional as you can completely test everything without the lock and it is quite complicated to set up. For this, you'll need an Adafruit electronic lock, a 1K Ohm resistor, and a power transistor.

Finally, here is the list of all the components that we will use in this project:

- Arduino Uno (`https://www.sparkfun.com/products/11021`)
- Adafruit Fona 808 shield (`http://www.adafruit.com/product/2542`)
- GSM uFL antenna (`http://www.adafruit.com/products/1991`)
- GSM SIM card
- 3.7V LiPo battery (`http://www.adafruit.com/products/328`)
- LiPo battery charger (`http://www.adafruit.com/products/1904`)
- LED (`https://www.sparkfun.com/products/9590`)
- 330 Ohm resistor (`https://www.sparkfun.com/products/8377`)
- 5V relay (`https://www.pololu.com/product/2480`)
- Optional: Adafruit electrical lock (`http://www.adafruit.com/products/1512`)
- Optional: Rectifier diode (`https://www.sparkfun.com/products/8589`)
- Optional: Power transistor (`http://www.adafruit.com/products/976`)
- Optional: 1K Ohm resistor (`https://www.sparkfun.com/products/8980`)
- Optional: 12V power supply (`https://www.sparkfun.com/products/9442`)
- Optional: DC jack adapter (`https://www.sparkfun.com/products/10288`)
- Breadboard (`https://www.sparkfun.com/products/12002`)
- Jumper wires (`https://www.sparkfun.com/products/8431`)

On the software side, you'll only need the latest version of the Arduino IDE and the Adafruit FONA library. You can install this library using the Arduino IDE library manager.

Hardware configuration

It's now the time to assemble the hardware for the project. We'll first connect the Adafruit FONA shield and then the other components. Before assembling the hardware, make sure that you have inserted the SIM card into the FONA shield.

The following is a schematic to help you out:

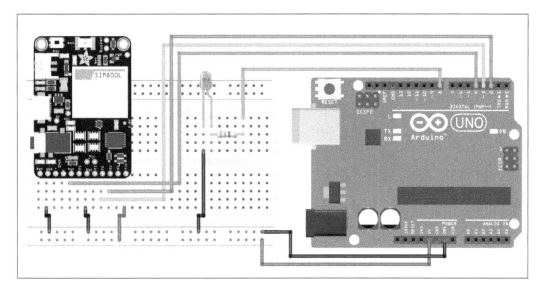

Note that the location of the pins can be different on your FONA shield, as there are many versions available. Also, note that the relay is not represented on this schematic. Here are the steps that you need to take in order to assemble the hardware:

1. First, connect the power supply to the breadboard. Connect the 5V pin from the Arduino board to the red power line on the breadboard and the GND pin to the blue power line.

2. Then, place the FONA shield on the breadboard. Connect the VIO pin to the red power line and the GND and key pins to the blue power line.

3. After that, connect the RST pin to Arduino pin 4, TX to Arduino pin 3, and RX to Arduino pin 2. Also, connect the 3.7V LiPo battery and antenna to the FONA shield.

4. Finally, connect the relay and the LED to the Arduino board. Connect the relay VCC and GND pin to the power supply on the breadboard and the EN pin to pin 7 of the Arduino board.

5. For the LED, place it in series with a 330 Ohm resistor on the breadboard, with the longest side of the LED connected to the resistor. Then, connect the other pin of the resistor to pin 8 of the Arduino board. Also, connect the other pin of the LED to the ground.

The following is an image of the completed project, with a zoom around the FONA shield and Arduino board:

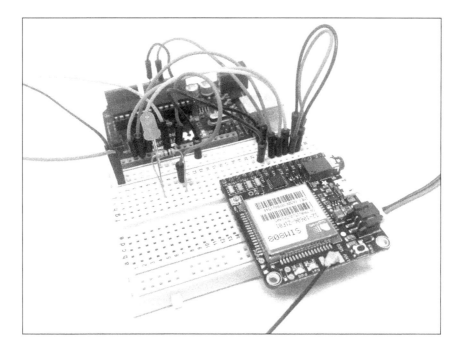

The following is an overview of the completely assembled project:

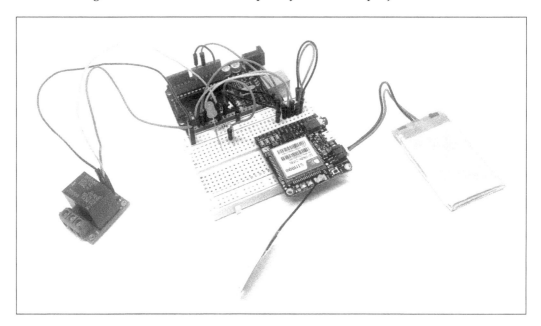

Testing the FONA shield

Now that our hardware is ready, we are going to test the FONA shield to see whether it is correctly connected and can connect to the network. As a test, we will send an SMS to the shield and also read all the messages present in the SIM card.

This is the complete Arduino sketch for this part, minus some helper functions that won't be detailed here:

```
// Include library
#include "Adafruit_FONA.h"
#include <SoftwareSerial.h>

// Pins
#define FONA_RX 2
#define FONA_TX 3
#define FONA_RST 4

// Buffer for replies
char replybuffer[255];

// Software serial
SoftwareSerial fonaSS = SoftwareSerial(FONA_TX, FONA_RX);
SoftwareSerial *fonaSerial = &fonaSS;

// Fona
Adafruit_FONA fona = Adafruit_FONA(FONA_RST);

// Readline function
uint8_t readline(char *buff, uint8_t maxbuff, uint16_t timeout = 0);

void setup() {

  // Start Serial
  while (!Serial);
  Serial.begin(115200);
  Serial.println(F("FONA basic test"));
  Serial.println(F("Initializing....(May take 3 seconds)"));

  // Init FONA
  fonaSerial->begin(4800);
  if (! fona.begin(*fonaSerial)) {
    Serial.println(F("Couldn't find FONA"));
    while(1);
  }
```

```
    Serial.println(F("FONA is OK"));

    // Print SIM card IMEI number.
    char imei[15] = {0}; // MUST use a 16 character buffer for IMEI!
    uint8_t imeiLen = fona.getIMEI(imei);
    if (imeiLen > 0) {
      Serial.print("SIM card IMEI: "); Serial.println(imei);
    }

}

void loop() {

  // Get number of SMS
  int8_t smsnum = fona.getNumSMS();
  if (smsnum < 0) {
    Serial.println(F("Could not read # SMS"));
  } else {
    Serial.print(smsnum);
    Serial.println(F(" SMS's on SIM card!"));
  }

  // Read last SMS
  flushSerial();
  Serial.print(F("Read #"));
  uint8_t smsn = smsnum;
  Serial.print(F("\n\rReading SMS #")); Serial.println(smsn);

  // Retrieve SMS sender address/phone number.
  if (! fona.getSMSSender(smsn, replybuffer, 250)) {
    Serial.println("Failed!");
  }
  Serial.print(F("FROM: ")); Serial.println(replybuffer);

  // Retrieve SMS value.
  uint16_t smslen;
  if (! fona.readSMS(smsn, replybuffer, 250, &smslen)) { // pass in
buffer and max len!
    Serial.println("Failed!");
  }
  Serial.print(F("***** SMS #")); Serial.print(smsn);
  Serial.print(" ("); Serial.print(smslen); Serial.println(F(") bytes
*****"));
  Serial.println(replybuffer);
```

```
    Serial.println(F("*****"));

    // Flush input
    flushSerial();
    while (fona.available()) {
      Serial.write(fona.read());
    }

    // Wait
    delay(10000);

  }
```

Now, let's see the details of this sketch. It starts by including the required libraries:

```
#include "Adafruit_FONA.h"
#include <SoftwareSerial.h>
```

Then, we will define the pins on which the FONA shield is connected:

```
#define FONA_RX 2
#define FONA_TX 3
#define FONA_RST 4
```

We will also define a buffer that will contain the SMS that we will read from the shield:

```
char replybuffer[255];
```

Then, we'll use a SoftwareSerial instance to communicate with the shield:

```
SoftwareSerial fonaSS = SoftwareSerial(FONA_TX, FONA_RX);
SoftwareSerial *fonaSerial = andfonaSS;
```

We will also create an instance of the FONA library:

```
Adafruit_FONA fona = Adafruit_FONA(FONA_RST);
```

Then, in the setup() function of the sketch, we will start the Serial object for the purpose of debugging and print a welcome message:

```
while (!Serial);
Serial.begin(115200);
Serial.println(F("FONA basic test"));
Serial.println(F("Initializing....(May take 3 seconds)"));
```

After that, we will initialize the FONA shield:

```
fonaSerial->begin(4800);
if (! fona.begin(*fonaSerial)) {
  Serial.println(F("Couldn't find FONA"));
  while(1);
}
Serial.println(F("FONA is OK"));
```

Then, we will read the IMEI of the SIM card in order to check whether the shield is working and whether a card is present:

```
char imei[15] = {0}; // MUST use a 16 character buffer for IMEI!
uint8_t imeiLen = fona.getIMEI(imei);
if (imeiLen > 0) {
  Serial.print("SIM card IMEI: "); Serial.println(imei);
}
```

Now, in the `loop()` function of the sketch, we will first count the number of text messages present in the card:

```
int8_t smsnum = fona.getNumSMS();
if (smsnum < 0) {
  Serial.println(F("Could not read # SMS"));
} else {
  Serial.print(smsnum);
  Serial.println(F(" SMS's on SIM card!"));
}
```

Then, we will also read the last one and print it on the serial monitor:

```
// Read last SMS
flushSerial();
Serial.print(F("Read #"));
uint8_t smsn = smsnum;
Serial.print(F("\n\rReading SMS #")); Serial.println(smsn);

 // Retrieve SMS sender address/phone number.
if (! fona.getSMSSender(smsn, replybuffer, 250)) {
  Serial.println("Failed!");
}
Serial.print(F("FROM: ")); Serial.println(replybuffer);

// Retrieve SMS value.
uint16_t smslen;
```

```
if (! fona.readSMS(smsn, replybuffer, 250, &smslen)) { // pass in
buffer and max len!
  Serial.println("Failed!");
}
Serial.print(F("***** SMS #")); Serial.print(smsn);
Serial.print(" ("); Serial.print(smslen); Serial.println(F(") bytes
*****"));
Serial.println(replybuffer);
Serial.println(F("*****"));
```

Once that's done, we will flush the `SoftwareSerial` instance and wait for 10 seconds until the next read:

```
// Flush input
flushSerial();
while (fona.available()) {
  Serial.write(fona.read());
}

// Wait
delay(10000);
```

 Note that the complete code for this section can be found on the GitHub repository of the book at `https://github.com/marcoschwartz/arduino-secret-agents`.

It's now the time to test the FONA shield. Make sure that you get all the code and open it in your Arduino IDE. Then, upload the code to the board and open the serial monitor. You should see the following screenshot:

```
                          /dev/cu.usbmodem1a12121
|                                                                    Send

FONA basic test
Initializing....(May take 3 seconds)
FONA is OK
SIM card IMEI: 865067020421544
8 SMS's on SIM card!
Read #
Reading SMS #8
FROM: +48733382390
***** SMS #8 (62) bytes *****
This is a test message for the Arduino for secret agents book.
*****
8 SMS's on SIM card!
Read #
Reading SMS #8
FROM: +48733382390
***** SMS #8 (62) bytes *****
This is a test message for the Arduino for secret agents book.
*****

✓ Autoscroll                          Carriage return  :    115200 baud  :
```

If everything is working fine, you should see the IMEI number of the SIM card and then the latest SMS in the card. You can test it by sending a message to the phone number of the SIM card; it should immediately show in the next reading of the SIM card. If this works, congratulations! You can now move to the next section.

Controlling the relay

In this part, we are going to program the Arduino board to remotely control the relay and LED that are connected to the board by sending an SMS to the FONA shield. As most of the code is similar to the code that we saw in the previous section, I'll only detail the new elements of code here.

We need to define the pins on which the relay and the LED are connected:

```
#define RELAY_PIN 7
#define LED_PIN 8
```

Then, we'll define a set of variables for a counter for the relay/lock. This is needed because after we open a lock, for example, we will want it to close automatically after a given amount of time for the purpose of safety. Here, we use a delay of 5 seconds:

```
bool lock_state = false;
int init_counter = 0;
int lock_counter = 0;
int lock_delay = 5000;
```

Then, we will define variables to count the number of SMS stored in the card:

```
int8_t smsnum = 0;
int8_t smsnum_old = 0;
```

We will also set two "passwords" to activate or deactivate the relay. I used very simple passphrases here; however, you can use the one that you want:

```
String password_on = "open";
String password_off = "close";
```

Then, in the `setup()` function of the sketch, we will set the relay and LED pins as output:

```
pinMode(RELAY_PIN, OUTPUT);
pinMode(LED_PIN, OUTPUT);
```

In the `loop()` function of the sketch, we will get the total number of SMS stored in the SIM card:

```
smsnum = fona.getNumSMS();
```

Basically, what we want is to check whether a new SMS has been received by the shield. To do that, we will compare this new reading to the old number of SMS stored in the SIM card:

```
if (smsnum > smsnum_old) {
```

If that's the case, we store the message in a `String` object so that we can conduct some operations on it:

```
String message = String(replybuffer);
```

Then, if the received message matches with the "On" password that we set earlier, we will activate the relay and LED, and also start the counter:

```
if (message.indexOf(password_on) > -1) {
  Serial.println("Lock OPEN");
  digitalWrite(RELAY_PIN, HIGH);
  digitalWrite(LED_PIN, HIGH);
  lock_state = true;
  init_counter = millis();
}
```

We would do something similar if we received an "Off" message:

```
if (message.indexOf(password_off) > -1) {
    Serial.println("Lock CLOSE");
    digitalWrite(RELAY_PIN, LOW);
    digitalWrite(LED_PIN, LOW);
    lock_state = false;
  }
```

After that, we will update the counter:

```
lock_counter = millis();
```

Then, we check whether the delay has already passed since the last time the relay/lock was activated. If that's the case, we will deactivate the relay/close the lock automatically:

```
if (lock_state == true andand (lock_counter - init_counter) > lock_
delay) {
  Serial.println("Lock CLOSE");
  digitalWrite(RELAY_PIN, LOW);
  digitalWrite(LED_PIN, LOW);
  lock_state = false;
}
```

Finally, we will store the current number of received text messages in the smsnum_old variable:

```
smsnum_old = smsnum;
```

Note that the complete code can be found in the GitHub repository of the project at https://github.com/marcoschwartz/arduino-secret-agents.

It's time to test the project. Grab all the code and paste it in the Arduino IDE. Make sure that you change the on/off passwords in the code as well.

Then, upload the code to the board and open the serial monitor. You should see the shield being initialized and then waiting for a new text message. Now, send a text message to the shield with the message corresponding to the "On" password that you defined earlier. The following screenshot is what you should see:

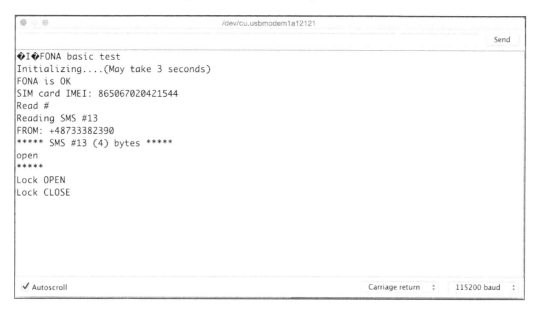

You should also see the LED turn on and the relay being activated. You can either wait until the system deactivates itself or you can simply send the "Off" passphrase again.

Opening and closing the lock

Now that we managed to actually control the relay and the LED by sending text messages to the FONA shield, we can connect the actual lock. This last step of the project is optional as you already tested the main functionality of the project and connecting the electronic door lock is a bit more technical. This is an image of the lock that I used:

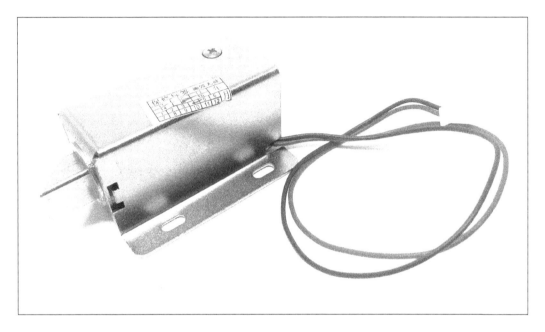

First, you'll need to cut the JST connector from the electronic lock in order to be able to connect it the breadboard. I soldered some 2-pin header at the end of the electronic lock to make it compatible with my breadboard.

Then, you'll need to assemble the remaining optional components on the breadboard, as shown in the following schematic:

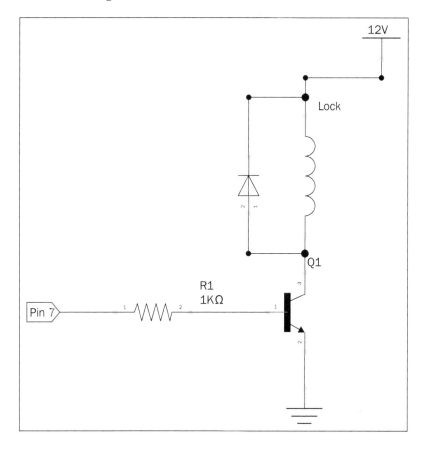

Let's see how to assemble these components together. First, place all the components on the breadboard. Then, connect the digital pin 7 of the Arduino board to the base of the transistor via the 1K Ohm resistor. Also, connect the emitter of the transistor to the ground of the project. Then, connect the electronic door lock between the collector of the transistor and the 12V power supply. Finally, mount the rectifier diode parallel to the door lock, with the orientation defined on the schematic.

Then, it's time to test the project again. You can use the exact same code as in the previous section. Congratulations, you can now control a door lock by sending a secret passphrase via SMS!

Summary

In this chapter, we built another very useful project for secret agents; a project where you can open a door lock by sending a secret code by SMS. You can also use this project for many other purposes. For example, you could use the same project to remotely activate an alarm when a text message is received by the project.

Note that you can also use the FONA shield to actually send text messages, which opens the door to even more exciting projects!

In the next chapter, we are going to see how to build a typical secret agent project: a remote spy camera that can be monitored from anywhere in the world!

6

Building a Cloud Spy Camera

We are now going to build a very famous project for secret agents: a spy camera. This will be a camera that can be, for example, installed in a room behind a set of books and help you in monitoring the room from another location.

We are going to build two projects based on the same hardware. These will be the key topics covered in this chapter:

- The first project will be a spy camera that takes a picture every time motion is detected in front of it and uploads it to your Dropbox account. The pictures will then be accessible to the spy from anywhere in the world.

- Finally, we'll end the chapter by making the camera stream live videos on a local Wi-Fi network. This will be perfect for a spy who wants to see what's happening in a room while being hidden in another room or outside. Let's dive in!

Hardware and software requirements

First, let's see what are the required components for this project.

For once, we are not going to use an Arduino Uno board, but an Arduino Yun. Not only do we need Wi-Fi connectivity but also the on-board USB port of the Yun. This will make it really easy to use a USB camera with our project.

The following is the Arduino board that I used for this project:

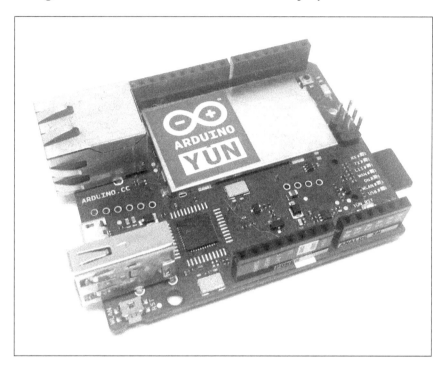

Then, you need a USB camera. You need a camera that is compatible with the **USB Video Class** (**UVC**). Basically, most recent USB cameras are compatible with this standard. I recommend the Logitech C270 USB camera that I used for this project:

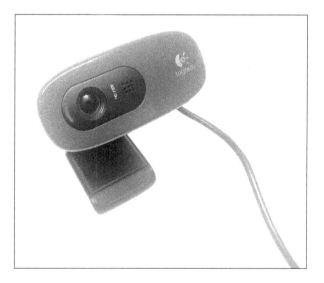

Finally, you will also need a PIR motion sensor, to detect whether there is motion in front of the camera. Any brand will be fine, you just need it be 5V-level compatible. This is the sensor that I used for this project:

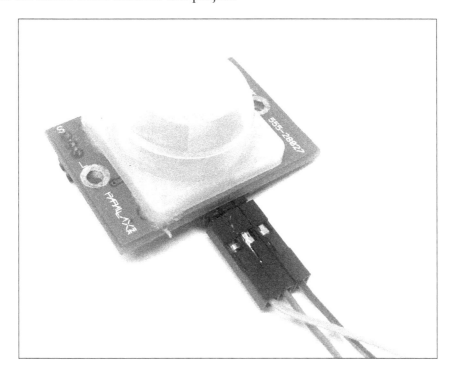

Finally, here is the list of all the components that we will use in this project:

- Arduino Yun (http://www.adafruit.com/product/1498)
- USB camera (http://www.logitech.com/en-us/product/hd-webcam-c270)
- PIR motion sensor (http://www.adafruit.com/product/189)
- A microSD card (at least 2 GB)
- Breadboard (https://www.sparkfun.com/products/12002)
- Jumper wires (https://www.sparkfun.com/products/8431)

On the software side, you will only need the Arduino IDE. If you are using Windows, you'll also need a terminal software. I recommend using PuTTY, which you can download from http://www.putty.org/.

Hardware configuration

Now let's configure the hardware for this project. It will be really simple and we'll also set up the Wi-Fi connectivity of Yun.

This is a schematic to help you out (excluding the USB camera):

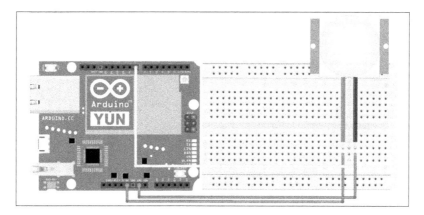

You just need to connect the PIR motion sensor to Yun. Connect VCC to the Arduino 5V pin, GND to GND, and the output of the sensor to pin number 8 of the Arduino Yun.

Finally, insert the USB camera into the USB port and the microSD card into the Arduino Yun.

The following is how it should look in the end:

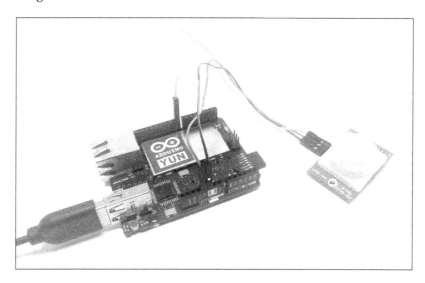

Now, we are going to set up Yun so that it connects to your Wi-Fi network. For that, the best way is to follow the latest instructions from Arduino that are available at `https://www.arduino.cc/en/Guide/ArduinoYun`.

Then, you should be able to connect to your Yun via your favorite web browser and access it with the password that you set earlier:

After that, you'll be able to see that your Yun is working:

Now, we'll access it from a terminal in order to install some modules. If you are on Windows, I recommend using PuTTY to type these commands. Start by connecting your Yun (replace the address with the one of your Yun) with:

```
ssh root@arduinoyun.local
```

You will then be greeted by the following message:

Now, type the following command:

```
opkg update
```

When this is done, type:

```
opkg install kmod-video-uvc
```

Then enter:

```
opkg install fswebcam
```

Finally, type the following command:

```
opkg install mjpg-streamer
```

Congratulations, your Yun is now fully operational for this project!

Setting up your Dropbox account

It's now time to set up your Dropbox account. First, make sure that you actually have an account by simply visiting the Dropbox website. Then, we'll need to create a Dropbox app. For this, go to https://www.dropbox.com/developers/apps.

Then, you can create a new app with the **Create app** button:

Give it a name and make sure it is set similar to the following screenshot:

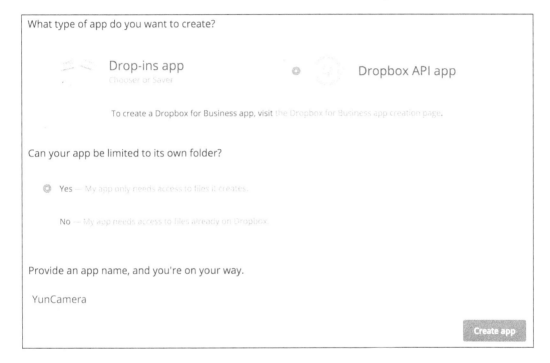

Now, in the parameters of the app, there are two things that you need: **App key** and **App secret**. You can find them both on the same page:

YunCameraSecret

Settings	Details	App metrics

Status	Development		Apply for production
Development users	Only you		Enable additional users
Permission type	App folder		
App folder name	YunCameraSecret		Change
App key	uth2bek7kunns95		
App secret	Show		

Once you have these, you can move to the next step and configure your Temboo account.

Setting up your Temboo account

We are going to use the Temboo service to link our hardware to the Dropbox app that we just created. This will allow us to upload files to Dropbox.

You first need to set up a new Temboo account from the following URL: `https://www.temboo.com/library/`.

Then, we need to actually authorize our Temboo account (and therefore, our Arduino project) in order to use your Dropbox app. For this, go to `https://www.temboo.com/library/Library/Dropbox/OAuth/InitializeOAuth/`.

You will be asked to enter your Dropbox App key and App secret:

Once you click on **Run**, there are two things that you'll need to do. First, you need to follow the link that is given to you by Temboo:

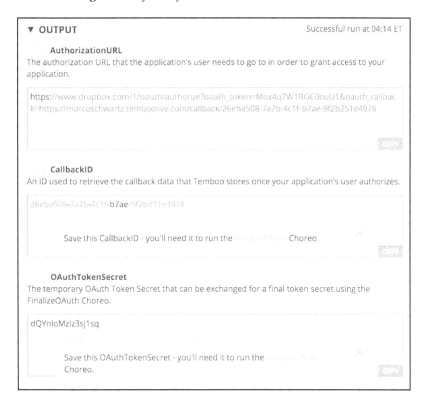

After that, you'll need to grab **CallbackID** and **OAuthTokenSecret** and go to the page at `https://www.temboo.com/library/Library/Dropbox/OAuth/FinalizeOAuth/`.

On this page, you can enter all the information that you have received so far:

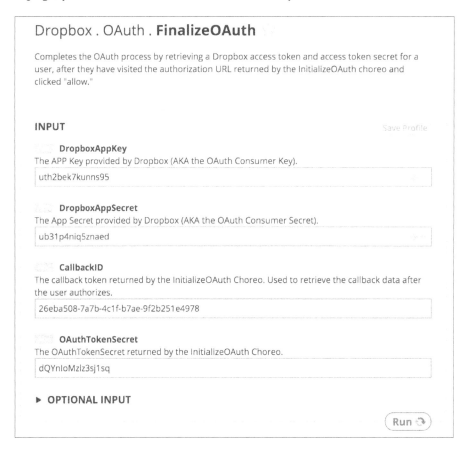

You'll be then given access token and token secret, which you will need for the spy camera project:

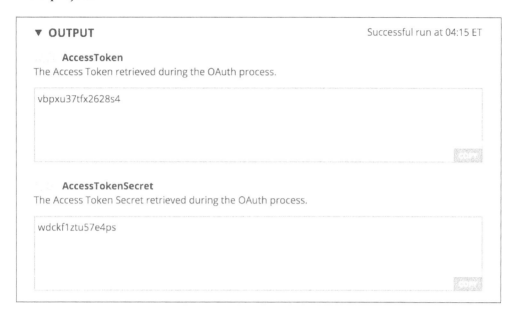

 Note that the Dropbox API is subject to change in the future. Therefore, always check the Temboo page and follow the instructions given there if they are different from the ones presented in this book.

You just need one more thing from Temboo; some data about your account. Go to `https://www.temboo.com/account/applications/`.

There you can see the information about the **Application** that you created when opening your account:

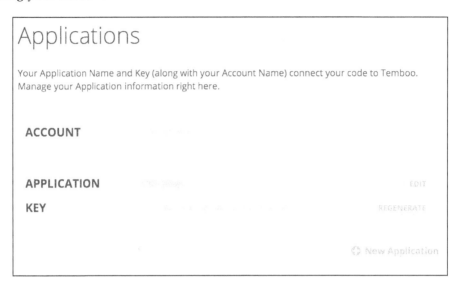

Keep this open, you'll also need this information later.

Saving pictures to Dropbox

Finally, we are going to make our first application using the hardware that we built. There will be two parts here: an Arduino sketch and a Python script. The Arduino sketch will be in charge of taking a picture in case motion is detected and call the Python script. The Python script will actually upload the pictures to Temboo every time it is called by the Arduino sketch.

This is the complete Arduino sketch:

```
// Sketch to upload pictures to Dropbox when motion is detected
#include <Bridge.h>
#include <Process.h>

// Picture process
Process picture;

// Filename
String filename;

// Pin
```

```
int pir_pin = 8;

// Path
String path = "/mnt/sda1/";

void setup() {

  // Bridge
  Bridge.begin();

  // Set pin mode
  pinMode(pir_pin,INPUT);
}

void loop(void)
{

  if (digitalRead(pir_pin) == true) {

    // Generate filename with timestamp
    filename = "";
    picture.runShellCommand("date +%s");
    while(picture.running());

    while (picture.available()>0) {
      char c = picture.read();
      filename += c;
    }
    filename.trim();
    filename += ".png";

    // Take picture
    picture.runShellCommand("fswebcam " + path + filename + " -r
1280x720");
    while(picture.running());

    // Upload to Dropbox
    picture.runShellCommand("python " + path + "upload_picture.py " +
path + filename);
    while(picture.running());
  }
}
```

Let's see what the most important parts of this sketch are. First, you need to include the required libraries:

```
#include <Bridge.h>
#include <Process.h>
```

Then, we will define the path to the SD card, which is where the pictures will be stored:

```
String path = "/mnt/sda1/";
```

After that, we will initialize the `Bridge` instance that will allow us to use the Yun filesystem, for example:

```
Bridge.begin();
```

Still in the `setup()` function of the sketch, we will set the motion sensor pin as an input:

```
pinMode(pir_pin, INPUT);
```

After that, in the `loop()` function of the sketch, we will check whether the motion sensor detected motion:

```
if (digitalRead(pir_pin) == true) {
```

If that's the case, we will first build a filename for the new picture using the current date and time:

```
filename = "";
    picture.runShellCommand("date +%s");
while(picture.running());

while (picture.available()>0) {
    char c = picture.read();
  filename += c;
}
filename.trim();
filename += ".png";
```

Then, we will use the `fswebcam` utility to save this picture on the SD card:

```
picture.runShellCommand("fswebcam " + path + filename + " -r
1280x720");
while(picture.running());
```

And finally, we will call the Python script to actually upload the picture to Dropbox:

```
picture.runShellCommand("python " + path + "upload_picture.py " + path
+ filename);
while(picture.running());
```

Now, let's see the Python script. The following is the complete script:

```
# coding=utf-8
# Script to upload files to Dropbox

# Import correct libraries
import base64
import sys
from temboo.core.session import TembooSession
from temboo.Library.Dropbox.FilesAndMetadata import UploadFile

print str(sys.argv[1])

# Encode image
with open(str(sys.argv[1]), "rb") as image_file:
    encoded_string = base64.b64encode(image_file.read())

# Declare Temboo session and Choreo to upload files
session = TembooSession('yourSession', 'yourApp', 'yourKey')
uploadFileChoreo = UploadFile(session)

# Get an InputSet object for the choreo
uploadFileInputs = uploadFileChoreo.new_input_set()

# Set inputs
uploadFileInputs.set_AppSecret("yourAppSecret")
uploadFileInputs.set_AccessToken("yourAccessToken")
uploadFileInputs.set_FileName(str(sys.argv[1]))
uploadFileInputs.set_AccessTokenSecret("yourTokenSecret")
uploadFileInputs.set_AppKey("yourAppKey")
uploadFileInputs.set_FileContents(encoded_string)
uploadFileInputs.set_Root("sandbox")

# Execute choreo
uploadFileResults = uploadFileChoreo.execute_with_
results(uploadFileInputs)
```

Now, let's see the most important parts of this script. We will first import the Temboo libraries:

```
import base64
import sys
from temboo.core.session import TembooSession
from temboo.Library.Dropbox.FilesAndMetadata import UploadFile
```

You will also need to set up your Temboo account details:

```
session = TembooSession('yourSession', 'yourApp', 'yourKey')
uploadFileChoreo = UploadFile(session)
```

Then, we will create a new set of inputs for the Dropbox library:

```
uploadFileInputs = uploadFileChoreo.new_input_set()
```

After that, this is where you will need to enter all the keys that we got from Dropbox and Temboo:

```
uploadFileInputs.set_AppSecret("yourAppSecret")
uploadFileInputs.set_AccessToken("yourAccessToken")
uploadFileInputs.set_FileName(str(sys.argv[1]))
uploadFileInputs.set_AccessTokenSecret("yourTokenSecret")
uploadFileInputs.set_AppKey("yourAppKey")
uploadFileInputs.set_FileContents(encoded_string)
uploadFileInputs.set_Root("sandbox")
```

Finally, we will execute the uploading of the file to Dropbox:

```
uploadFileResults = uploadFileChoreo.execute_with_
results(uploadFileInputs)
```

It's now the time to test the project! Note that you can find all the code from the GitHub repository at https://github.com/marcoschwartz/arduino-secret-agents.

You will still need to download the Temboo Python library from https://temboo.com/sdk/python.

Then, make sure that you modified the Python files with your own data. Also, put the SD card in your computer again using an adapter. Place the file and the `temboo` SDK on the SD card, as shown in the following screenshot:

After that, place the SD card back in your Yun. Open the Arduino sketch with the Arduino IDE and make sure that you have selected the Arduino Yun board. Now, upload the sketch to the board.

Now, try to move your hand in front of the motion sensor; the sensor should go red and you should also notice that the camera becomes active immediately (there is a little green LED on the Logitech C270 camera).

You can also check your Dropbox account in the `Applications` folder. There should be a new folder that has been created, which contains the pictures taken by the spy camera:

Congratulations, you now built your first spy camera project! Note that this folder can, of course, be accessed from anywhere in the world so even if you are on the other side of the town, you can monitor what's going on in the room where the camera is sitting.

Live streaming from the spy camera

We are now going to end this chapter with a shorter project: using the camera to stream live video in a web browser. This stream will be accessible from any device connected to the same Wi-Fi network as Yun.

To start with this project, log in to your Yun using the following command (by changing the name of the board with the name of your Yun):

```
ssh root@arduinoyun.local
```

Then, type the following:

```
mjpg_streamer -i "input_uvc.so -d /dev/video0 -r 640x480 -f 25" -o
"output_http.so -p 8080 -w /www/webcam" &
```

This will start the streaming from your Yun. You can now simply go the URL of your Yun, and add `:8080` at the end, for example, `http://arduinoyun.local:8080`.

You will arrive at the streaming interface:

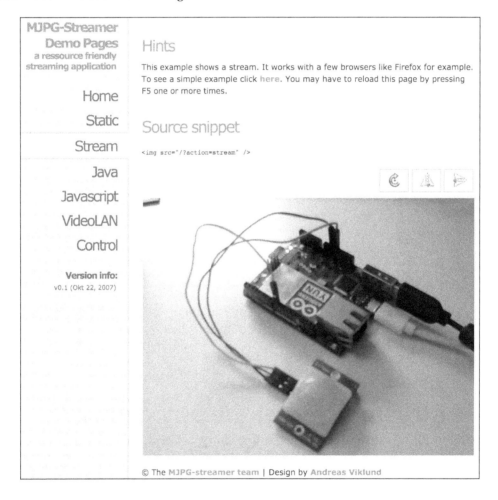

You can now stream this video live to your mobile phone or any other device in the same network. It's the perfect project to spy on a room while you are sitting outside, for example.

Summary

In this project, we built a spy camera project that can send pictures to the Cloud whenever motion is detected. We also saw that it can do other things such as streaming live videos in a web browser.

There are, of course, many ways to improve this project. You can, for example, deploy several of these spy cameras in a building and make them take pictures that you identify in the Arduino code (when the filename is created).

In the next chapter, we are going to make another exciting application for secret agents: a project that will allow you to monitor secret data from anywhere in the world!

7
Monitoring Secret Data from Anywhere

In this chapter, we are going to build a project that will continuously record data from sensors, and send this data over Wi-Fi so it's accessible from any web browser. This is great for a secret agent that wants to monitor a room remotely, without being seen. You'll of course be able to adapt the project with your own sensors, depending on what you want to record.

To do so, these are the steps we are going to take in this chapter:

- We will use Arduino along with the CC3000 Wi-Fi chip, which is quite convenient to give Wi-Fi connectivity to Arduino projects.
- We will send sensor data to an online service called dweet.io, and then display the result on a dashboard using Freeboard.io.
- Finally, we'll also see how to set up automated alerts based on the recorded data. Let's dive in!

Hardware and software requirements

First, let's see what the required components are for this project.

We'll of course use an Arduino Uno as the brain of the project. For Wi-Fi connectivity, we are going to use a CC3000 breakout board from Adafruit:

We'll also use a bunch of sensors to illustrate the behavior of the project: a DHT11 sensor for temperature and humidity, a photocell for light levels, and a motion sensor.

Finally, here is a list of all the components that we will use in this project:

- Arduino Uno (https://www.sparkfun.com/products/11021)
- CC3000 breakout board (http://www.adafruit.com/product/1469)
- DHT11 sensor with 4.7k Ohm resistor (http://www.adafruit.com/product/386)
- Photocell (https://www.sparkfun.com/products/9088)
- 10k Ohm resistor (https://www.sparkfun.com/products/8374)
- PIR motion sensor (http://www.adafruit.com/product/189)
- Breadboard (https://www.sparkfun.com/products/12002)
- Jumper wires (https://www.sparkfun.com/products/8431)

On the software side, you need the latest version of the Arduino IDE. You will also need the following libraries:

- Adafruit CC3000 library
- Adafruit DHT library

To install these libraries, just use the Arduino library manager.

Hardware configuration

Now let's assemble the different components of this project. This is a schematic to help you out:

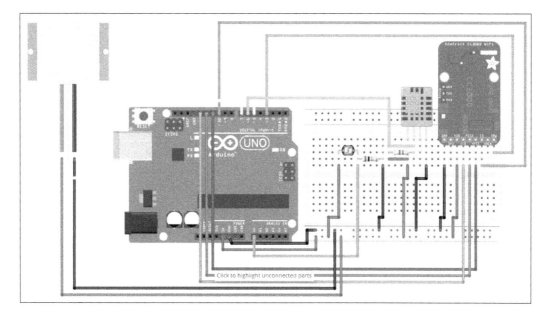

First, connect the power. Connect the Arduino Uno 5V to the red power rail on the breadboard, and the GND pin to the blue power rail. Also, place all the main components on the breadboard.

After that, for the DHT11 sensor, follow the instructions given by the schematic to connect the sensor to the Arduino board. Make sure you don't forget the 4.7k Ohm resistor between the VCC and signal pins.

We are now going to connect the photocell. Start by placing the photocell on the breadboard in series with the 10k Ohm resistor. After that, connect the other end of the photocell to the red power rail, and the other pin of the resistor to the blue power rail. Finally, connect the pin between the photocell and the resistor to Arduino Uno pin A0.

Finally, for the motion sensor, connect the VCC pin to the red power rail, the GND pin to the blue power rail, and finally the output pin of the sensor to Arduino pin 7.

Now, we are going to connect the CC3000 breakout board. Connect the pins as indicated on the schematic: IRQ to pin number 3 of the Arduino board, VBAT to pin 5, and CS to pin 10. After that, connect these pins to the Arduino board: MOSI, MISO, and CLK go to pins 11, 12, and 13, respectively. Finally, connect the power to the CC3000 breakout: connect 5V to the Vin pin of the breakout board, and GND to GND.

This is a picture of the completely assembled project:

Congratulations, your project is now fully assembled! You can move on to the next part: sending data to the cloud.

Sending data to dweet.io

The first step in this project is really to send data to the web so it is stored online. For this, we'll use a service called dweet.io. You can check it out at `https://dweet.io/`.

This is the main welcome page:

This is the complete Arduino code for this project:

```
// Libraries
#include <Adafruit_CC3000.h>
#include <SPI.h>
#include "DHT.h"
#include <avr/wdt.h>

// Define CC3000 chip pins
#define ADAFRUIT_CC3000_IRQ    3
#define ADAFRUIT_CC3000_VBAT   5
#define ADAFRUIT_CC3000_CS     10

// DHT sensor
#define DHTPIN 6
#define DHTTYPE DHT11

// Create CC3000 instances
Adafruit_CC3000 cc3000 = Adafruit_CC3000(ADAFRUIT_CC3000_CS, ADAFRUIT_
CC3000_IRQ, ADAFRUIT_CC3000_VBAT,
```

```
                                          SPI_CLOCK_DIV2); // you can
change this clock speed

// DHT instance
DHT dht(DHTPIN, DHTTYPE);

// WLAN parameters
#define WLAN_SSID        "yourWiFiSSID"
#define WLAN_PASS        "yourWiFiPassword"
#define WLAN_SECURITY    WLAN_SEC_WPA2

// Dweet parameters
#define thing_name  "mySecretThing"

// Variables to be sent
int temperature;
int humidity;
int light;
int motion;

uint32_t ip;

void setup(void)
{
  // Initialize
  Serial.begin(115200);

  Serial.println(F("\nInitializing..."));
  if (!cc3000.begin())
  {
    Serial.println(F("Couldn't begin()! Check your wiring?"));
    while(1);
  }

  // Connect to WiFi network
  Serial.print(F("Connecting to WiFi network ..."));
  cc3000.connectToAP(WLAN_SSID, WLAN_PASS, WLAN_SECURITY);
  Serial.println(F("done!"));

  /* Wait for DHCP to complete */
  Serial.println(F("Request DHCP"));
  while (!cc3000.checkDHCP())
  {
    delay(100);
```

```
  }

  // Start watchdog
  wdt_enable(WDTO_8S);
}

void loop(void)
{

  // Measure from DHT
  float t = dht.readTemperature();
  float h = dht.readHumidity();
  temperature = (int)t;
  humidity = (int)h;

  // Measure light level
  float sensor_reading = analogRead(A0);
  light = (int)(sensor_reading/1024*100);

  // Get motion sensor reading
  motion = digitalRead(7);
  Serial.println(F("Measurements done"));

  // Reset watchdog
  wdt_reset();

  // Get IP
  uint32_t ip = 0;
  Serial.print(F("www.dweet.io -> "));
  while (ip == 0) {
    if (! cc3000.getHostByName("www.dweet.io", &ip)) {
      Serial.println(F("Couldn't resolve!"));
    }
    delay(500);
  }
  cc3000.printIPdotsRev(ip);
  Serial.println(F(""));

  // Reset watchdog
  wdt_reset();

  // Check connection to WiFi
  Serial.print(F("Checking WiFi connection ..."));
  if(!cc3000.checkConnected()){while(1){}}
```

```
  Serial.println(F("done."));
  wdt_reset();

  // Send request
  Adafruit_CC3000_Client client = cc3000.connectTCP(ip, 80);
  if (client.connected()) {
    Serial.print(F("Sending request... "));

    client.fastrprint(F("GET /dweet/for/"));
    client.print(thing_name);
    client.fastrprint(F("?temperature="));
    client.print(temperature);
    client.fastrprint(F("&humidity="));
    client.print(humidity);
    client.fastrprint(F("&light="));
    client.print(light);
    client.fastrprint(F("&motion="));
    client.print(motion);
    client.fastrprintln(F(" HTTP/1.1"));

    client.fastrprintln(F("Host: dweet.io"));
    client.fastrprintln(F("Connection: close"));
    client.fastrprintln(F(""));

    Serial.println(F("done."));
  } else {
    Serial.println(F("Connection failed"));
    return;
  }

  // Reset watchdog
  wdt_reset();

  Serial.println(F("Reading answer..."));
  while (client.connected()) {
    while (client.available()) {
      char c = client.read();
      Serial.print(c);
    }
  }
  Serial.println(F(""));

  // Reset watchdog
```

```
wdt_reset();

// Close connection and disconnect
client.close();
Serial.println(F("Closing connection"));
Serial.println(F(""));

// Reset watchdog & disable
wdt_reset();

}
```

Now let's look at the most important parts of the code. First, we need to include the required libraries, such as the CC3000 library and the DHT library:

```
#include <Adafruit_CC3000.h>
#include <SPI.h>
#include "DHT.h"
#include <avr/wdt.h>
```

Then, we define which pin the DHT11 sensor is connected to:

```
#define DHTPIN 6
#define DHTTYPE DHT11
```

You also need to enter your Wi-Fi name and password:

```
#define WLAN_SSID       "yourWiFiSSID"
#define WLAN_PASS       "yourWiFiPassword"
#define WLAN_SECURITY   WLAN_SEC_WPA2
```

Then, you can define a name for your thing that is, the virtual object that will store the data online:

```
#define thing_name   "mySecretThing"
```

In the setup() function of the sketch, we initialize the CC3000 chip:

```
if (!cc3000.begin())
  {
    Serial.println(F("Couldn't begin()! Check your wiring?"));
    while(1);
  }
```

We also connect to the Wi-Fi network:

```
cc3000.connectToAP(WLAN_SSID, WLAN_PASS, WLAN_SECURITY);
```

Finally, we initialize the watchdog to 8 seconds. This will basically reset the Arduino automatically if we don't refresh it before this delay. It is basically here to prevent the project from getting stuck:

```
wdt_enable(WDTO_8S);
```

In the `loop()` function of the sketch, we first measure data from the DHT sensor:

```
float t = dht.readTemperature();
float h = dht.readHumidity();
temperature = (int)t;
humidity = (int)h;
```

After that, we measure the ambient light level:

```
float sensor_reading = analogRead(A0);
light = (int)(sensor_reading/1024*100);
```

And finally we get the status of the motion sensor:

```
motion = digitalRead(7);
```

Then, we try to get the IP address of the dweet.io website:

```
while (ip == 0) {
  if (! cc3000.getHostByName("www.dweet.io", &ip)) {
    Serial.println(F("Couldn't resolve!"));
  }
  delay(500);
}
```

Then, we connect the project to this IP address:

```
Adafruit_CC3000_Client client = cc3000.connectTCP(ip, 80);
```

We can now send the data, following the format given by dweet.io:

```
client.fastrprint(F("GET /dweet/for/"));
client.print(thing_name);
client.fastrprint(F("?temperature="));
client.print(temperature);
client.fastrprint(F("&humidity="));
client.print(humidity);
client.fastrprint(F("&light="));
client.print(light);
client.fastrprint(F("&motion="));
client.print(motion);
client.fastrprintln(F(" HTTP/1.1"));
```

After that, we read the answer from the server:

```
while (client.connected()) {
    while (client.available()) {
        char c = client.read();
        Serial.print(c);
    }
}
```

And we close the connection:

```
client.close();
```

It's now time to test the project! Make sure you grab all the code, copy it inside the IDE, and change the Wi-Fi details and thing name. Then, upload the code to the board and open the Serial monitor:

```
                                    /dev/cu.usbmodem1a12121
                                                                          Send

Initializing...
Connecting to WiFi network ...done!
Request DHCP
Measurements done
www.dweet.io -> 54.172.56.193
Checking WiFi connection ...done.
Sending request... done.
Reading answer...
HTTP/1.1 200 OK
Access-Control-Allow-Origin: *
Content-Type: application/json
Content-Length: 185
Date: Thu, 03 Sep 2015 09:38:07 GMT
Connection: close

{"this":"succeeded","by":"dweeting","the":"dweet","with":{"thing":"mySecretThing","created":"2015-09-0:
Closing connection

  Autoscroll                              Carriage return  :     115200 baud  :
```

You should see that the project is sending measurements to dweet.io and then getting an answer. The most important part is the one indicating that data was recorded:

```
{"this":"succeeded","by":"dweeting","the":"dweet","with":{"thing":"myS
ecretThing","created":"2015-09-03T09:38:07.051Z","content":{"temperatu
re":28,"humidity":32,"light":87,"motion":0}}}
```

You can also check online to make sure data was recorded:

← → C 🔒 https://**dweet.io**/get/latest/dweet/for/mySecretThing

{"this":"succeeded","by":"getting","the":"dweets","with":[{"thing":"mySecretThing","created":"2015-09-03T09:40:48.782Z","content":
{"temperature":28,"humidity":32,"light":88,"motion":1}}]}

Now that we are sure that data is being recorded, we can move to the next step: spying on this data remotely from any web browser!

Monitoring the device remotely

We are now going to see how to access the data stored on dweet.io and display it graphically. For that, we are going to use a website called freeboard.io. You can go there with the following URL `http://freeboard.io/`.

This is the main welcome screen where you need to create an account:

Once you have an account, you can create a new board:

Once this is done, you should be redirected to a similar page showing an empty board:

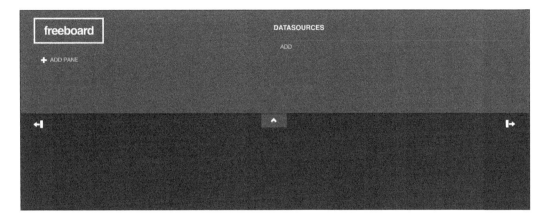

First, we need to set a datasource, meaning that we need to tell Freeboard to get data from the dweet thing we are storing the data in. Add a new datasource and fill out the fields as shown in the following screenshot, with the name of the thing that stores your data of course:

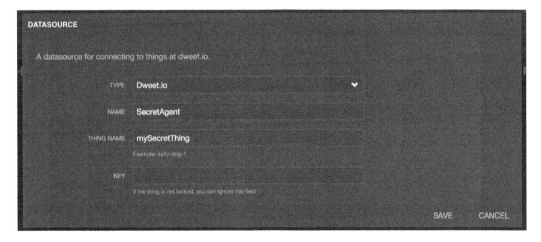

After that, you will see the datasource appearing at top of your board, with the last update date:

It's now time to add some graphical elements to our dashboard. We'll first add one for the temperature. Click on a new pane, which will create a new block inside the dashboard:

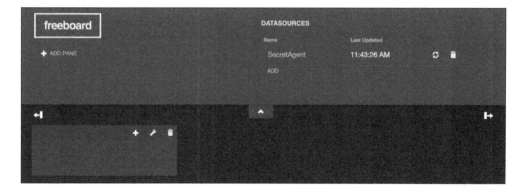

Then, click on the little **+** sign to create a new widget. Here, we are going to use a gauge widget for the temperature. Fill out the widget creation form as in the following screenshot:

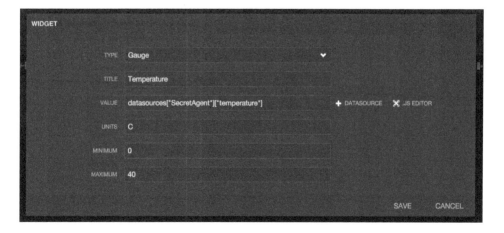

You should immediately see the gauge for the temperature on your board:

Now, let's do the same for humidity:

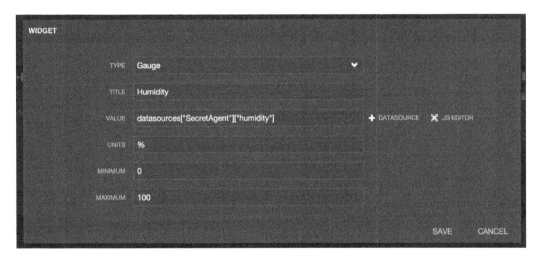

You can also do the same for the ambient light level. You now have all the data from these sensors refreshed in near real-time on your board:

The last thing we need to put in is the motion sensor. As it's an on/off sensor, I used an indicator widget for this:

Now try to pass your hand in front of the sensor. You should immediately see that the indicator changes its color:

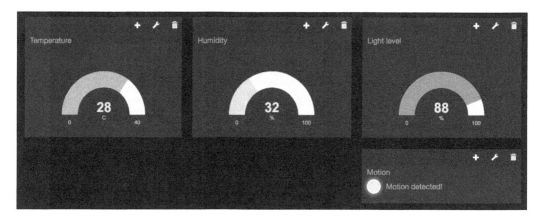

Congratulations, you now have a dashboard that you can access at any time to spy on this data!

Creating automated e-mail alerts

There is one more thing we can do with our project. Spying on the data is good, but we are not always behind a computer. For example, we would like to receive an alert via e-mail if motion is detected by the project.

Dweet.io proposes this kind of service, at a very cheap price (less than $1 a month). To do this, you need to make your device private with a lock. This is basically a key you can get from `https://dweet.io/locks`.

This is the screen where you can buy this key:

Once we have the key, we can actually set our alert. The dweet.io website explains this very well:

To set up an alert, simply go to the following URL by modifying the different parameters with the desired parameters `https://dweet.io/alert/youremail@` `yourdomain.com/when/yourThing/dweet.motion==1?key=yourKey`.

Once that's done, you will automatically receive an alert whenever motion is detected by your project, even if you are not actually watching the data!

Summary

In this chapter, we built a project that can just be put in a room to spy on it from anywhere in the world. It continuously records data from the sensors connected to it, and sends this data directly to the cloud.

There are of course many things you can do to improve this project. You can, for example, experiment with your own sensors, depending on the data you want to spy on. You can also try to replace the CC3000 Wi-Fi chip with the GSM/GPRS shield we used in an earlier chapter to have a module you can place somewhere without having to connect it to a Wi-Fi network.

In the next chapter, we'll dive into a more advanced topic: actually building a GPS tracker based on Arduino!

8
Creating a GPS Tracker with Arduino

We are now going to build a very common tool that any secret agent should have: a GPS tracker. We'll use the Arduino platform again to build our own DIY GPS tracker. We will actually build two applications using the same hardware:

- The first one will be a location device that sends its position via SMS

- The other project will be a GPS tracker that you can actually locate live on a map

Let's dive in!

Hardware and software requirements

First let's see what the required components are for this project. As usual, we are going to use an Arduino Uno board as the central part of the project.

For the GPRS and GPS parts, we are going to use the Adafruit FONA 808 module again, which we already used in *Chapter 5, Opening a Lock with an SMS*. We'll also need a GSM antenna for GSM/GPRS communications.

However, as here we want to use the onboard GPS, we'll also need an additional GPS antenna. I used a standard uFL passive GPS antenna from Adafruit:

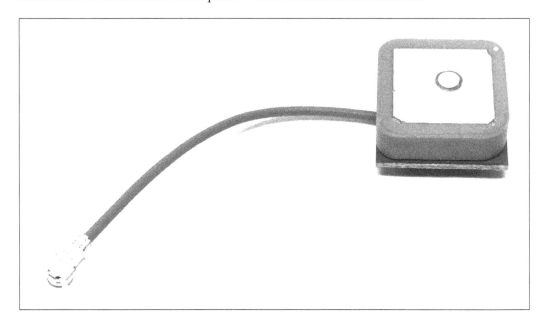

Then, you will need a battery to power the FONA shield and the onboard GPS module, as the Arduino Uno board doesn't allow you to power the FONA shield (it can use up to 2A at a time!). For that, I used a 3.7V LiPo battery, along with a microUSB battery charger.

A very important part of the project is the SIM card, which you need to place inside the FONA shield. You will need a normal SIM card (not micro or nano), which is activated, not locked by a PIN, and able to receive text messages. You can get one at any of your local mobile network operators. Also, in this chapter you also need to have a data plan active with at least 1 MB of credit on the account.

Finally, here is a list of all the components that we will use in this project:

- Arduino Uno (https://www.sparkfun.com/products/11021)
- Adafruit Fona 808 breakout (http://www.adafruit.com/product/2542)
- GSM uFL antenna (http://www.adafruit.com/products/1991)
- GSM SIM card
- 3.7V LiPo battery (http://www.adafruit.com/products/328)
- LiPo battery charger (http://www.adafruit.com/products/1904)
- Passive GPS antenna (https://www.adafruit.com/product/2461)

- Breadboard (https://www.sparkfun.com/products/12002)
- Jumper wires (https://www.sparkfun.com/products/8431)

On the software side, you'll only need the latest version of the Arduino IDE, and the Adafruit FONA library. You can install this library using the Arduino IDE library manager.

Hardware configuration

It's now time to assemble the hardware of this project.

First, connect the power supply to the breadboard: connect the 5V pin from the Arduino board to the red power line on the breadboard, and the GND pin to the blue power line.

Then, place the FONA shield on the breadboard. Connect the VIO pin to the red power line, and the GND and Key pins to the blue power line.

After that, connect the RST pin to Arduino pin 4, TX to Arduino pin 3, and RX to Arduino pin 2. Also connect the 3.7V LiPo battery, the GPS antenna, and the GSM antenna to the FONA shield.

This is a close-up picture of the shield after the project was assembled:

And this is an overview of the whole project assembled:

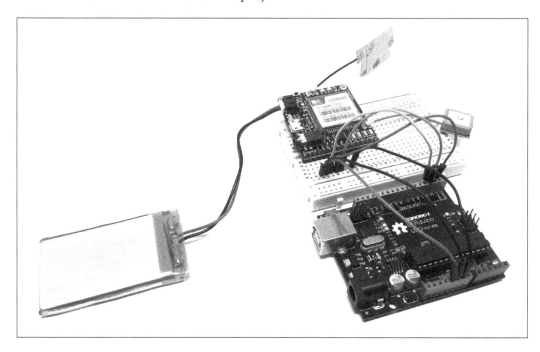

Testing the location functions

Before we dive into the two exciting projects of this chapter, we'll first make a simple test using the FONA shield, and see whether it can actually locate our project on a map. The sketch will actually test whether the GPS location is working correctly.

If not, no worries: the sketch, and the other projects of this chapter, will automatically use GPRS location instead. It's less precise, but works quite well. This will be the case if you are testing this project inside, for example.

This is the complete code for this part:

```
// Libraries
#include "Adafruit_FONA.h"
#include <SoftwareSerial.h>

// Pins
#define FONA_RX 2
#define FONA_TX 3
```

```
#define FONA_RST 4

// Buffer
char replybuffer[255];

// Instances
SoftwareSerial fonaSS = SoftwareSerial(FONA_TX, FONA_RX);
SoftwareSerial *fonaSerial = &fonaSS;
Adafruit_FONA fona = Adafruit_FONA(FONA_RST);

void setup() {

  // Init board
  while (!Serial);
  Serial.begin(115200);
  Serial.println(F("FONA location test"));
  Serial.println(F("Initializing....(May take 3 seconds)"));

  fonaSerial->begin(4800);
  if (! fona.begin(*fonaSerial)) {
    Serial.println(F("Couldn't find FONA"));
    while(1);
  }
  Serial.println(F("FONA is OK"));

  // Print SIM card IMEI number.
  char imei[15] = {0}; // MUST use a 16 character buffer for IMEI!
  uint8_t imeiLen = fona.getIMEI(imei);
  if (imeiLen > 0) {
    Serial.print("SIM card IMEI: "); Serial.println(imei);
  }

  // Setup GPRS APN (username/password optional)
  fona.setGPRSNetworkSettings(F("your_APN"));
  //fona.setGPRSNetworkSettings(F("your_APN"), F("your_username"),
F("your_password"));

  // Turn GPS on
  if (!fona.enableGPS(true)) {
    Serial.println(F("Failed to turn on GPS"));
  }

  // Turn GPRS on
```

```
  fona.enableGPRS(true);

  // Decide between GPS or GPRS localisation
  boolean GPSloc;
  int8_t stat;

  // Check GPS fix
  stat = fona.GPSstatus();
  if (stat < 0) {
    GPSloc = false;
  }
  if (stat == 0 || stat == 1) {
    GPSloc = false;
  }
  if (stat == 2 || stat == 3) {
    GPSloc = false;
  }

  // Print which localisation method is used
  Serial.print("Localisation method: ");
  if (GPSloc) {Serial.println("GPS");}
  else {Serial.println("GPRS");}

  // Print position
  if (GPSloc) {
    char gpsdata[80];
    fona.getGPS(0, gpsdata, 80);
    Serial.println(F("Reply in format: mode, longitude, latitude,
altitude, utctime(yyyymmddHHMMSS), ttff, satellites, speed, course"));
    Serial.println(gpsdata);
  }
  else {
    uint16_t returncode;
    if (!fona.getGSMLoc(&returncode, replybuffer, 250))
      Serial.println(F("Failed!"));
    if (returncode == 0) {
      Serial.println(replybuffer);
    } else {
      Serial.print(F("Fail code #")); Serial.println(returncode);
    }
  }
}

void loop() {
  // Nothing here
}
```

We are now going to look at the important parts of this sketch.

It starts by importing the required libraries:

```
#include "Adafruit_FONA.h"
#include <SoftwareSerial.h>
```

Then, we define the pins that the FONA is connected to:

```
#define FONA_RX 2
#define FONA_TX 3
#define FONA_RST 4
```

We create some instances for the `SoftwareSerial` object and the FONA breakout:

```
SoftwareSerial fonaSS = SoftwareSerial(FONA_TX, FONA_RX);
SoftwareSerial *fonaSerial = &fonaSS;
Adafruit_FONA fona = Adafruit_FONA(FONA_RST);
```

Then, we initialize the FONA:

```
while (!Serial);
Serial.begin(115200);
Serial.println(F("FONA location test"));
Serial.println(F("Initializing....(May take 3 seconds)"));

fonaSerial->begin(4800);
if (! fona.begin(*fonaSerial)) {
  Serial.println(F("Couldn't find FONA"));
  while(1);
}
Serial.println(F("FONA is OK"));
```

Following this is where you need to put your GPRS data. This completely depends on your phone carrier. I simply had to put in an APN, which was internet, but you might have to put a username and password as well. Contact your carrier to get the exact information, and then comment/uncomment the required line and fill out the data:

```
fona.setGPRSNetworkSettings(F("your_APN"));
//fona.setGPRSNetworkSettings(F("your_APN"), F("your_username"),
F("your_password"));
```

Then, we can enable the GPS:

```
if (!fona.enableGPS(true)) {
  Serial.println(F("Failed to turn on GPS"));
}
```

We also enable the GPRS connection:

```
fona.enableGPRS(true);
```

When this is done, we get the state of the GPS, and check whether we can locate a GPS satellite and get a fix. If yes, we'll use the GPS for location purposes. If not, we'll switch to a GPRS location:

```
stat = fona.GPSstatus();
if (stat < 0) {
  GPSloc = false;
}
if (stat == 0 || stat == 1) {
  GPSloc = false;
}
if (stat == 2 || stat == 3) {
  GPSloc = false;
}
```

Once we know which location method to use, we can actually get the location, depending on the chosen method:

```
if (GPSloc) {
  location = getLocationGPS();
  latitude = getLatitudeGPS(location);
  longitude = getLongitudeGPS(location);
  latitudeNumeric = convertDegMinToDecDeg(latitude.toFloat());
  longitudeNumeric = convertDegMinToDecDeg(longitude.toFloat());
}
else {
  location = getLocationGPRS();
  latitude = getLatitudeGPRS(location);
  longitude = getLongitudeGPRS(location);
  latitudeNumeric = latitude.toFloat();
  longitudeNumeric = longitude.toFloat();
}
Serial.print("Latitude, longitude: ");
Serial.print(latitudeNumeric, 6);
Serial.print(",");
Serial.println(longitudeNumeric, 6);
```

As you can see, we are calling a few helper functions here to get the `latitude` and `longitude` as float variables. We are now going to see all these functions in detail.

Here is the detail of the function to get the location using GPS:

```
String getLocationGPS() {

  // Buffer
  char gpsdata[80];

  // Get data
  fona.getGPS(0, gpsdata, 80);
  return String(gpsdata);
}
```

This function is really simple, as it simply gets the data from the GPS of the FONA, and returns it as a `String`.

The function to get the location using GPRS is similar:

```
String getLocationGPRS() {

  // Buffer for reply and returncode
  char replybuffer[255];
  uint16_t returncode;

  // Get and return location
  if (!fona.getGSMLoc(&returncode, replybuffer, 250))
    return String("Failed!");
  if (returncode == 0) {
    return String(replybuffer);
  } else {
    return String(returncode);
  }

}
```

Then, we actually need to extract the data from the returned String object, and get the latitude and longitude of our GPS module. Here is the function for the longitude using the GPRS location:

```
String getLongitudeGPRS(String data) {

  // Find commas
  int commaIndex = data.indexOf(',');
  int secondCommaIndex = data.indexOf(',', commaIndex+1);

  return data.substring(0, commaIndex);
}
```

This is the function to get the latitude using the GPRS location:

```
String getLatitudeGPRS(String data) {

  // Find commas
  int commaIndex = data.indexOf(',');
  int secondCommaIndex = data.indexOf(',', commaIndex+1);

  return data.substring(commaIndex + 1, secondCommaIndex);
}
```

For the GPS location, things are similar. However, we need one extra function for the GPS module. Indeed, the latitude and longitude are given in degrees-minutes-seconds, and we need to convert this to a numeric format. This is done using the following function:

```
double convertDegMinToDecDeg (float degMin) {
  double min = 0.0;
  double decDeg = 0.0;

  // Get the minutes, fmod() requires double
  min = fmod((double)degMin, 100.0);

  //rebuild coordinates in decimal degrees
  degMin = (int) ( degMin / 100 );
  decDeg = degMin + ( min / 60 );

  return decDeg;
}
```

It's now time to test the sketch! Simply put all the code in the Arduino IDE, update your GPRS settings, and upload it to the board. Then, open the serial monitor and this is what you should see:

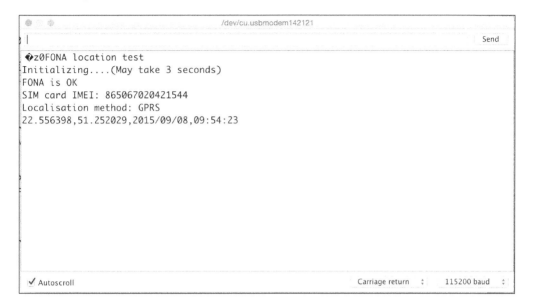

As you can see, I was testing this while indoors so the sketch used the GPRS location. I then simply copied the latitude and longitude and pasted them into Google Maps:

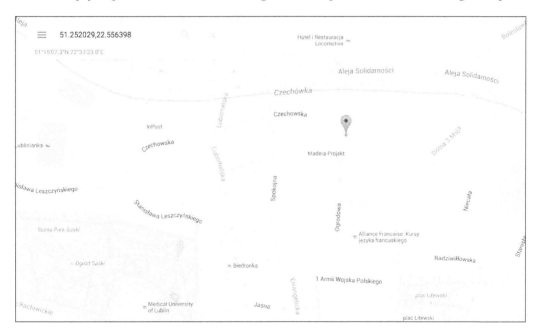

I immediately saw a point close to my own location, which was 100-150 meters off my real location. Not too bad if you need to track whether something is inside or outside a city for example.

I then tried again outside, with nothing between the GPS antenna and the sky. Then, the sketch automatically selected the GPS location option:

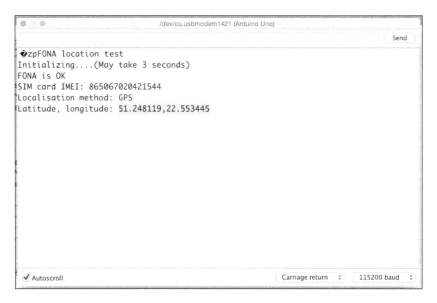

I then also copied and pasted the GPS latitude and longitude into Google Maps:

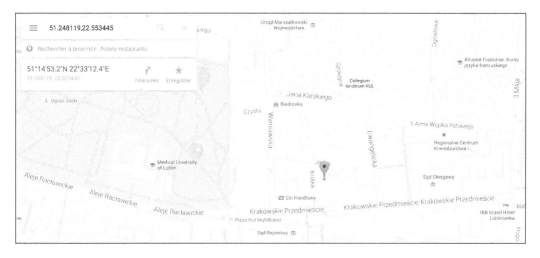

This time, the location was really accurate and showed exactly where I was at that moment.

Sending a GPS location by SMS

Now that we have the location part working correctly, we can start building our GPS tracking projects for secret agents. The first project will simply use the location and send it via SMS to the phone number of your choice.

As the sketch is quite similar to previous one, I'll only show which parts changed compared to the location test sketch. We first need to define the phone number where you want to send the tracking data to:

```
char * sendToNumber = "123456789";
```

After getting the location just as in the previous section, we can now build the message that we will send via SMS:

```
char messageToSend[140];
String message = "Current GPS location: " + latitude + "," +
longitude;
message.toCharArray(messageToSend, message.length());
```

Using the FONA instance, it's very easy to actually send the SMS to the number we defined earlier:

```
if (!fona.sendSMS(sendToNumber, messageToSend)) {
  Serial.println(F("Failed to send SMS"));
} else {
  Serial.println(F("Sent location data!"));
}
```

Finally, we wait before each message. I used 10 seconds for testing purposes, but you can enter your own delay in milliseconds:

```
Serial.println(F("Waiting until next message..."));
delay(10000);
```

It's now time to test the project. You can grab the whole code from the GitHub repository of the book at https://github.com/marcoschwartz/arduino-secret-agents.

Then, make sure you change the phone number and your GPRS settings inside the code and upload it to the board. This is what you should see:

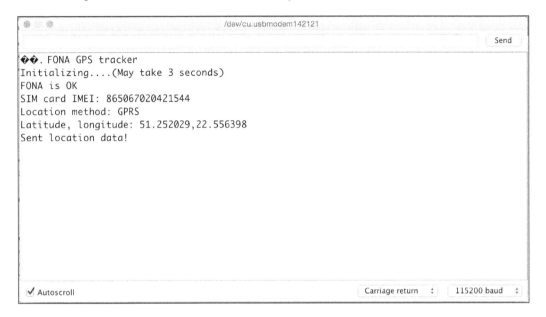

If you can see the last line, it means that an SMS actually has been sent. I quickly received the message after this:

This was followed by a series of messages with the current location of the project:

If you can see this message on your phone, congratulations, you have built your first GPS tracker using Arduino!

Building a GPS location tracker

It's now time to build the last project of this chapter: a real GPS location tracker. For this project, we'll get the location just as before, using the GPS if available, and the GPRS location otherwise.

However, here we are going to use the GPRS capabilities of the shield to send the latitude and longitude data to dweet.io, which is a service we have used before. Then, we'll plot this data in Google Maps, allowing you to follow your device in real time from anywhere in the world.

As the sketch is very similar to the ones in the previous sections, I'll only detail the most important parts of the sketch here.

You need to define a name for the thing that will contain the GPS location data:

```
String dweetThing = "mysecretgpstracker";
```

Then, after getting the current location, we prepare the data to be sent to dweet.io:

```
uint16_t statuscode;
int16_t length;
char url[80];
String request = "www.dweet.io/dweet/for/" + dweetThing + "?latitude="
+ latitude + "&longitude=" + longitude;
request.toCharArray(url, request.length());
```

After this, we actually send the data to dweet.io:

```
if (!fona.HTTP_GET_start(url, &statuscode, (uint16_t *)&length)) {
   Serial.println("Failed!");
 }
 while (length > 0) {
   while (fona.available()) {
     char c = fona.read();

     // Serial.write is too slow, we'll write directly to Serial
register!
     #if defined(__AVR_ATmega328P__) || defined(__AVR_ATmega168__)
       loop_until_bit_is_set(UCSR0A, UDRE0); /* Wait until data
register empty. */
       UDR0 = c;
     #else
       Serial.write(c);
     #endif
     length--;
   }
 }
 fona.HTTP_GET_end();
```

Now, before testing the project, we are going to prepare our dashboard that will host the Google Maps widget. Just as in the previous chapter, we are going to use freeboard.io for this purpose. If you don't have an account yet, go to `http://freeboard.io/`.

Create a new dashboard, and also a new datasource. Insert the name of your `thing` inside the **THING NAME** field:

Then, create a new pane with a Google Maps widget. Link this widget to the latitude and longitude of your location datasource:

It's now time to test the project. Make sure you grab all the code, for example from the GitHub repository of the book. Also, don't forget to modify the thing name, as well as your GPRS settings.

Then, upload the sketch to the board and open the Serial monitor. This is what you should see:

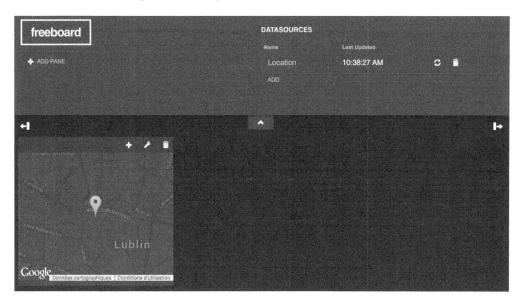

The most important line is the last one, which confirms that data has been sent to dweet.io and has been stored there.

Now, simply go back to the dashboard you just created, and you can now see that the location on the map has been updated:

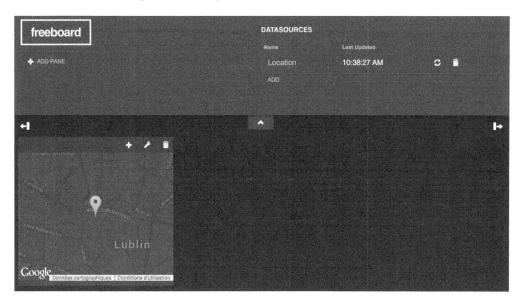

Note that this map is also updated in real time as new measurements arrive from the board. You can also modify the delay between two updates of the position of the tracker by changing the `delay()` function in the sketch. Congratulations, you just built your own GPS tracking device!

Summary

In this chapter, we built a device that allows us to track the GPS location of any object it is attached to. We built two projects based on the same hardware: one that sends SMS messages to your phone with the location of the device, and another one where you can see the current location of the device on a map.

There are of course many ways to improve this project. You can, for example, add sensors to the project and monitor them as well from the Freeboard dashboard, just as we did in the previous project. One useful thing is also to make the project autonomous, and for this, you can simply power the Arduino board directly from the FONA shield, as it has a 5V output.

In the final chapter of the book, we are going to build another cool application for secret agents: a small mobile surveillance robot.

Building an Arduino Spy Robot

In this last project of the book, we are going to build a small surveillance spy robot based on Arduino. The secret agent will be able to command the robot remotely from a web page and see what the robot is seeing in real time using a camera.

To do this, we will use everything you have learned in this book so far:

- How to use Wi-Fi with Arduino
- How to create control interfaces
- How to use a camera with the Arduino Yun

Let's dive in!

Hardware and software requirements

First let's see what the required components are for this project. The core of the project will of course be the robot itself. For the robot's chassis, you have a wide range of choices available on the market. For this project, I chose a DFRobot MiniQ 2 wheels chassis, which is a small robot chassis that you can easily mount Arduino boards to.

Then, you will need two Arduino boards for this project. The first one will be an Arduino Yun, which we will use to connect a USB camera, just as we did in *Chapter 6, Building a Cloud Spy Camera*. For the camera itself, I used a C720 camera from Logitech again.

The other thing you will need is an Arduino Uno, which will take care of driving the motors of the robot via a motor shield. We have to use an Arduino Uno here because the Arduino Yun is incompatible with most of the motor shields of DFRobot.

To control the motors, I used a 1A motor shield from DFRobot. In order to control the robot remotely, we'll also use the CC3000 breakout board that we used in an earlier chapter. To assemble the CC3000 board with the rest of the robot, we will also use a DFRobot prototyping shield.

You will also need a 7.2V battery pack to power the robot when it is not connected to your computer. I also used a DFRobot battery pack with a DC jack adapter.

Finally, here is a list of all the components that we will use in this project:

- DFRobot MiniQ 2 wheels chassis (`http://www.dfrobot.com/index.php?route=product/product&product_id=367#.VfP4kmR96u4`)
- DFRobot Motor shield (`http://www.dfrobot.com/index.php?route=product/product&product_id=59&search=motor+shield&description=true#.VfP4wWR96u4`)
- DFRobot Prototyping shield (`http://www.dfrobot.com/index.php?route=product/product&product_id=55&search=prototyping&description=true#.VfP40WR96u4`)
- CC3000 breakout board (`https://www.adafruit.com/products/1469`)
- Arduino Uno (`https://www.sparkfun.com/products/11021`)
- Arduino Yun (`http://www.adafruit.com/product/1498`)
- USB camera (`http://www.logitech.com/en-us/product/hd-webcam-c270`)
- Breadboard (`https://www.sparkfun.com/products/12002`)
- Jumper wires (`https://www.sparkfun.com/products/8431`)
- 7.2V battery with DC power jack (`http://www.dfrobot.com/index.php?route=product/product&product_id=489`)

On the software side, you will need the latest version of the Arduino IDE. You will also need to install the following libraries using the Arduino library manager:

- Adafruit CC3000 library
- aREST

Hardware configuration

Let's start with the most difficult part of this project: assembling the robot itself. Of course, the exact steps will depend on the robot chassis you are using, but the goal is to give you the main steps in this section.

We start by actually mounting the Arduino Yun board on the robot chassis using the screws and headers provided with the chassis itself:

Once that's done, we mount the Arduino Uno board on top of the Yun using an extra row of headers. Then, we tightly screw the Arduino Uno board into the metallic headers.

Then, we mount the motor shield on top of the Arduino Uno board. Next, we insert the cables from the motors into the dedicated headers on the motor shield and secure them with the screws. Make sure that you are using the same polarity for both motors.

Here is the result at this point:

Now, we are going to assemble the CC3000 breakout board on the prototyping shield, which will then sit on top of the robot. To know how to connect the CC3000 board, please refer to *Chapter 7, Monitoring Secret Data from Anywhere*. The only difference here is that the IRQ pin is connected to pin 3, and the VBAT pin is connected to pin 8.

This is the assembled CC3000 board with the prototyping shield:

Now, simply mount this shield on top of the robot:

Here is a side view of the project at this stage:

Finally, plug the camera into the USB port of the Arduino Yun and place it in front of the robot. I secured it with screws, but this will depend on your own robot chassis.

This is a close-up picture showing the completely assembled robot:

This is an overview of the completely assembled robot:

If you have a similar result, congratulations! You are now ready to program your surveillance robot. Don't worry about the battery for now, as we will do everything using USB cables that connect the robot directly to your computer.

Setting up the motor control

We are now going to set up the different parts of the project, starting with configuring the Arduino Uno, which will control the motors via Wi-Fi. For this, we'll use the aREST Arduino library, which makes it really easy to control Arduino projects via Wi-Fi.

Here is the complete code for this part:

```
// Robot test via aREST + WiFi
#define NUMBER_VARIABLES 1
#define NUMBER_FUNCTIONS 5

// Libraries
#include <Adafruit_CC3000.h>
#include <SPI.h>
#include <aREST.h>
#include <avr/wdt.h>

// CC3000 pins
#define ADAFRUIT_CC3000_IRQ     3
#define ADAFRUIT_CC3000_VBAT    8
#define ADAFRUIT_CC3000_CS      10

// Robot speed
#define FULL_SPEED 100
#define TURN_SPEED 50

// Motor pins
int speed_motor1 = 6;
int speed_motor2 = 5;
int direction_motor1 = 7;
int direction_motor2 = 4;

// Sensor pins
int distance_sensor = A0;

// CC3000 instance
Adafruit_CC3000 cc3000 = Adafruit_CC3000(ADAFRUIT_CC3000_CS, ADAFRUIT_
CC3000_IRQ, ADAFRUIT_CC3000_VBAT);

// Create aREST instance
aREST rest = aREST();

// The port to listen for incoming TCP connections
#define LISTEN_PORT             80

// Server instance
Adafruit_CC3000_Server restServer(LISTEN_PORT);

#define WLAN_SSID       "KrakowskiePrzedm51m.15(flat15)"
```

```
#define WLAN_PASS        "KrK51flat15_1944_15"
#define WLAN_SECURITY    WLAN_SEC_WPA2

void setup(void)
{
  // Start Serial
  Serial.begin(115200);

  // Give name to robot
  rest.set_id("1");
  rest.set_name("robot");

  // Expose functions
  rest.function("forward",forward);
  rest.function("backward",backward);
  rest.function("left",left);
  rest.function("right",right);
  rest.function("stop",stop);

  // Set up CC3000 and get connected to the wireless network
  Serial.print(F("Initialising CC3000..."));
  if (!cc3000.begin())
  {
    while(1);
  }
  Serial.println(F("done"));

  Serial.print(F("Connecting to WiFi..."));
  if (!cc3000.connectToAP(WLAN_SSID, WLAN_PASS, WLAN_SECURITY)) {
    while(1);
  }
  Serial.println(F("done"));

  Serial.print(F("Getting DHCP..."));
  while (!cc3000.checkDHCP())
  {
    delay(100);
  }

  // Start server
  restServer.begin();
```

```
    Serial.println(F("Listening for connections..."));

    displayConnectionDetails();

    wdt_enable(WDTO_8S);
}

void loop() {

    // Handle REST calls
    Adafruit_CC3000_ClientRef client = restServer.available();
    rest.handle(client);
    wdt_reset();

    // Check connection
    if(!cc3000.checkConnected()){while(1){}}
    wdt_reset();
}

// Forward
int forward(String command) {

    send_motor_command(speed_motor1,direction_motor1,100,1);
    send_motor_command(speed_motor2,direction_motor2,100,1);
    return 1;
}

// Backward
int backward(String command) {

    send_motor_command(speed_motor1,direction_motor1,100,0);
    send_motor_command(speed_motor2,direction_motor2,100,0);
    return 1;
}

// Left
int left(String command) {

    send_motor_command(speed_motor1,direction_motor1,75,0);
    send_motor_command(speed_motor2,direction_motor2,75,1);
    return 1;
}

// Right
```

```
int right(String command) {

  send_motor_command(speed_motor1,direction_motor1,75,1);
  send_motor_command(speed_motor2,direction_motor2,75,0);
  return 1;
}

// Stop
int stop(String command) {

  send_motor_command(speed_motor1,direction_motor1,0,1);
  send_motor_command(speed_motor2,direction_motor2,0,1);
  return 1;
}

// Function to command a given motor of the robot
void send_motor_command(int speed_pin, int direction_pin, int pwm,
boolean dir)
{
  analogWrite(speed_pin,pwm); // Set PWM control, 0 for stop, and 255
for maximum speed
  digitalWrite(direction_pin,dir);
}

// Print connection details of the CC3000 chip
bool displayConnectionDetails(void)
{
  uint32_t ipAddress, netmask, gateway, dhcpserv, dnsserv;

  if(!cc3000.getIPAddress(&ipAddress, &netmask, &gateway, &dhcpserv,
&dnsserv))
  {
    Serial.println(F("Unable to retrieve the IP Address!\r\n"));
    return false;
  }
  else
  {
    Serial.print(F("\nIP Addr: ")); cc3000.printIPdotsRev(ipAddress);
    Serial.print(F("\nNetmask: ")); cc3000.printIPdotsRev(netmask);
    Serial.print(F("\nGateway: ")); cc3000.printIPdotsRev(gateway);
    Serial.print(F("\nDHCPsrv: ")); cc3000.printIPdotsRev(dhcpserv);
    Serial.print(F("\nDNSserv: ")); cc3000.printIPdotsRev(dnsserv);
    Serial.println();
    return true;
  }
}
```

As this code is quite long, we are only going to look at the most important parts here. We start by including all the required libraries:

```
#include <Adafruit_CC3000.h>
#include <SPI.h>
#include <aREST.h>
#include <avr/wdt.h>
```

Then, we define the pins that correspond to the motor shield:

```
int speed_motor1 = 6;
int speed_motor2 = 5;
int direction_motor1 = 7;
int direction_motor2 = 4;
```

After this, you need to enter your own Wi-Fi network name and password:

```
#define WLAN_SSID       "your_wifi_ssid"
#define WLAN_PASS       "your_wifi_password"
#define WLAN_SECURITY   WLAN_SEC_WPA2s
```

Then, we declare the aREST instance, which we'll use later to access motor functions via Wi-Fi:

```
aREST rest = aREST();
```

In the setup() function of the sketch, we expose the different functions to control the robot so they are accessible via Wi-Fi:

```
rest.function("forward",forward);
rest.function("backward",backward);
rest.function("left",left);
rest.function("right",right);
rest.function("stop",stop);
```

After that, we start a Wi-Fi server on the board:

```
restServer.begin();
Serial.println(F("Listening for connections..."));
```

In the loop() function of the sketch, we continuously listen for connections and handle them with the aREST instance:

```
Adafruit_CC3000_ClientRef client = restServer.available();
rest.handle(client);
```

Now let's have a look at one of the functions to control the robot, for example, the one to go forward:

```
int forward(String command) {

    send_motor_command(speed_motor1,direction_motor1,100,1);
    send_motor_command(speed_motor2,direction_motor2,100,1);
    return 1;
}
```

As you can see, the work is done by another function that directly acts on the motors:

```
void send_motor_command(int speed_pin, int direction_pin, int pwm,
boolean dir)
{
    analogWrite(speed_pin,pwm);
    digitalWrite(direction_pin,dir);
}
```

It's now time to configure this part of the project. Connect the Arduino Uno to the computer via USB and upload this sketch to the board. Then, open the Serial monitor. After a while, you should see the IP address of the board being printed on the Serial monitor.

Setting up live streaming

We can now move to the next part: configuring live camera streaming on the Arduino Yun. This is something we already did in *Chapter 6, Building a Cloud Spy Camera*, so we'll only look at the most important parts here. Refer to the *Hardware configuration* section if you need to know how to configure your Yun again.

First, connect to your Yun using the following command:

ssh root@arduinoyun.local

Then, launch camera streaming with the following command:

mjpg_streamer -i "input_uvc.so -d /dev/video0 -r 640x480 -f 25" -o "output_http.so -p 8080 -w /www/webcam" &

You can check that the streaming is working at the following page http://arduinoyun.local:8080.

Setting up the interface

We can now move to the last part: setting up the interface that will allow a secret agent to command the robot and also to see the live stream from the camera. This interface will be composed of an HTML page and a JavaScript file. It will be based on the aREST.js module that makes it easy to control aREST devices from a web page.

This is the complete HTML page:

```
<!DOCTYPE html>
<html>
<head>
  <meta charset=utf-8 />
  <title>Surveillance Robot</title>
  <link rel="stylesheet" type="text/css" href="https://maxcdn.
bootstrapcdn.com/bootstrap/3.3.4/css/bootstrap.min.css">
  <link rel="stylesheet" type="text/css" href="interface.css">
  <script type="text/javascript" src="https://code.jquery.com/jquery-
2.1.4.min.js"></script>
  <script type="text/javascript" src="https://cdn.rawgit.com/Foliotek/
AjaxQ/master/ajaxq.js"></script>
  <script type="text/javascript" src="https://cdn.rawgit.com/
marcoschwartz/aREST.js/master/aREST.js"></script>
  <script type="text/javascript" src="interface.js"></script>
</head>

<body>

<div class='container'>

<h1>Surveillance Robot</h1>

<div class='row'>

  <div class="col-md-2"></div>

  <div class="col-md-2">
    <button id='fw' class='btn btn-primary btn-block'
type="button">Forward</button>
  </div>

</div>

<div class='row'>

  <div class="col-md-2">
```

```
    <button id='left' class='btn btn-primary btn-block'
type="button">Left</button>
  </div>

  <div class="col-md-2">
    <button id='stop' class='btn btn-danger btn-block'
type="button">Stop</button>
  </div>

  <div class="col-md-2">
    <button id='right' class='btn btn-primary btn-block'
type="button">Right</button>
  </div>

</div>

<div class='row'>

  <div class="col-md-2"></div>

  <div class="col-md-2">
    <button id='bw' class='btn btn-primary btn-block'
type="button">Backward</button>
  </div>

</div>

<div class='row'>

  <img src="http://arduinoyun.local:8080/?action=stream" />

</div>

</div>

</body>
</html>
```

The most important parts on this page are the buttons to control the robot and the live stream of the camera. For example, this defines the button to move the robot forward:

```
<div class="col-md-2">
    <button id='fw' class='btn btn-primary btn-block'
type="button">Forward</button>
  </div>
```

This `<img>` tag allows you to insert the live stream of the camera into the page:

```
<img src="http://arduinoyun.local:8080/?action=stream" />
```

Now let's look at the JavaScript file that will actually send the command to the robot. This is the complete file:

```
$( document ).ready(function() {

    // Device
    var address = '192.168.1.105';
    var device = new Device(address);

    // Button
    $('#fw').click(function() {
      device.callFunction('forward', '');
    });

    $('#bw').click(function() {
      device.callFunction('backward', '');
    });

    $('#left').click(function() {
      device.callFunction('left', '');
    });

    $('#right').click(function() {
      device.callFunction('right', '');
    });

    $('#stop').click(function() {
      device.callFunction('stop', '');
    });

});
```

You only have to change one thing in this script: the IP address of your board, which you obtained previously:

```
var address = '192.168.1.105';
var device = new Device(address);
```

The different lines of code on the page each control a function of the robot. For example, this piece of code creates the link between the forward button and the `forward` function on the robot:

```
$('#fw').click(function() {
    device.callFunction('forward', '');
});
```

Testing the surveillance robot

Now that we have all the elements in place, it's time to test our little surveillance robot. Note that you can grab the complete code from `https://github.com/marcoschwartz/arduino-secret-agents`.

Make sure that you configured the code and the hardware with the instructions from the previous sections.

Now, it's time to power your robot from the battery. Insert the battery into the robot chassis, and then plug it into the Arduino Uno DC jack input.

Then, connect the Arduino Uno 5V pin to the Vin pin of the Yun. Also connect the GND pins of the two boards together. This will ensure that the Arduino Yun will be powered as well. Don't forget to start the live video stream again after that.

Now, place your robot on the ground and open the HTML page. This is what you should see:

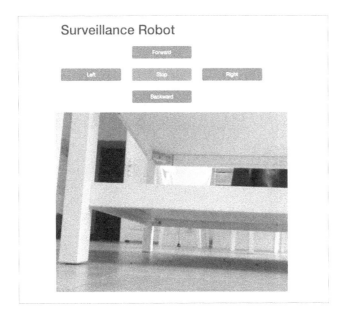

Now, try the different buttons, and you should see that the robot reacts nearly instantly. Congratulations, you just built your own spy robot based on Arduino!

Summary

In this last chapter of the book, you built your own little surveillance robot controlled by Wi-Fi. With this robot, you can now spy on what's going on in a room and also move the robot around.

There are several ways to improve this project. For example, you can add more sensors to the robot and display this sensor data on the same page that you control the robot from.

You have already reached the end of this book, and I hope that you enjoyed reading all the secret agent projects! I now really encourage you to try all the projects from this book again, and build them yourself. I also advise that you use all the knowledge from this book and build your own secret agent projects. Have fun!

Index

A

access control system, with fingerprint sensor
  access, controlling to relay 49-52
  components, URL 42
  fingerprint, enrolling 44-49
  GitHub repository, URL 43
  hardware, configuring 43, 44
  hardware requisites 41-43
  secret data, accessing 52-56
  software requisites 41-43
alarm system
  configuring 6-10
  GitHub repository, URL 10
  hardware, configuring 4-6
  hardware requisites 1-4
  software requisites 1-4
  testing 10, 11
automated e-mail alerts
  creating 112, 113

D

device
  monitoring, remotely 108-112
door lock
  closing 73, 74
  components, URL 62
  FONA shield, testing 65-70
  GitHub repository, URL 69
  hardware, configuring 62-64
  hardware requisites 60-62
  opening 73, 74
  relay, controlling 70- 72
  software requisites 60-62

Dropbox
  account, setting up 82-84
  pictures, saving 88-93
  URL 82
dweet.io
  data, sending 101-107
  URL 101, 113

E

EMF bug detector
  building 34-39
  GitHub repository, URL 38
  hardware, configuring 31, 32
  hardware requisites 29, 30
  LCD screen, testing 33, 34
  software requisites 29, 30

F

Freeboard.io
  URL 108

G

GitHub repository
  URL 130
GPS tracker
  building 129-132
  components, URL 116
  GitHub repository, URL 127
  GPS location, sending by SMS 127, 128
  hardware, configuring 117, 118
  hardware requisites 115-117
  location functions, testing 118-126
  software requisites 115-117

L

LCD screen
  testing, for EMF bug detector 33, 34

S

SD card
  spy microphone, recording on 25, 26
  used, for spy microphone 15-19
secret data monitor
  automated e-mail alerts, creating 112, 113
  components, URL 98
  data, sending to dweet.io 101-107
  device, monitoring remotely 108-112
  hardware, configuring 99, 100
  hardware requisites 98
  software requisites 98
spy camera
  components, URL 79
  Dropbox account, setting up 83, 84
  hardware, configuring 80-82
  hardware requisites 77-79
  live streaming 93-95
  pictures, saving to Dropbox 88-93
  software requisites 77-79
  Temboo account, setting up 84-88
spy microphone
  building 22-25

GitHub repository, URL 25
hardware requisites 13-15
recording, on SD card 25, 26
SD card, using 15-19
software requisites 13-15
testing 20-22
surveillance spy robot
  components, URL 134
  GitHub repository, URL 151
  hardware, configuring 134-141
  hardware requisites 133, 134
  interface, setting up 148-150
  live streaming, setting up 147
  motor control, setting up 141-147
  software requisites 133, 134
  testing 151, 152

T

Temboo
  account, setting up 84-88
  URL 84
Temboo Python library
  URL 92

U

USB Video Class (UVC) 78

Thank you for buying
Arduino for Secret Agents

About Packt Publishing

Packt, pronounced 'packed', published its first book, *Mastering phpMyAdmin for Effective MySQL Management*, in April 2004, and subsequently continued to specialize in publishing highly focused books on specific technologies and solutions.

Our books and publications share the experiences of your fellow IT professionals in adapting and customizing today's systems, applications, and frameworks. Our solution-based books give you the knowledge and power to customize the software and technologies you're using to get the job done. Packt books are more specific and less general than the IT books you have seen in the past. Our unique business model allows us to bring you more focused information, giving you more of what you need to know, and less of what you don't.

Packt is a modern yet unique publishing company that focuses on producing quality, cutting-edge books for communities of developers, administrators, and newbies alike. For more information, please visit our website at www.packtpub.com.

About Packt Open Source

In 2010, Packt launched two new brands, Packt Open Source and Packt Enterprise, in order to continue its focus on specialization. This book is part of the Packt Open Source brand, home to books published on software built around open source licenses, and offering information to anybody from advanced developers to budding web designers. The Open Source brand also runs Packt's Open Source Royalty Scheme, by which Packt gives a royalty to each open source project about whose software a book is sold.

Writing for Packt

We welcome all inquiries from people who are interested in authoring. Book proposals should be sent to author@packtpub.com. If your book idea is still at an early stage and you would like to discuss it first before writing a formal book proposal, then please contact us; one of our commissioning editors will get in touch with you.

We're not just looking for published authors; if you have strong technical skills but no writing experience, our experienced editors can help you develop a writing career, or simply get some additional reward for your expertise.

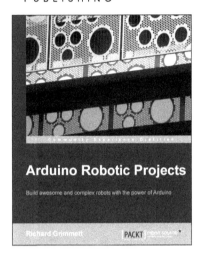

Arduino Robotic Projects

ISBN: 978-1-78398-982-9   Paperback: 240 pages

Build awesome and complex robots with the power of Arduino

1. Develop a series of exciting robots that can sail, go under water, and fly.

2. Simple, easy-to-understand instructions to program Arduino.

3. Effectively control the movements of all types of motors using Arduino.

4. Use sensors, GPS, and a magnetic compass to give your robot direction and make it lifelike.

Arduino Home Automation Projects

ISBN: 978-1-78398-606-4   Paperback: 132 pages

Automate your home using the powerful Arduino platform

1. Interface home automation components with Arduino.

2. Automate your projects to communicate wirelessly using XBee, Bluetooth and WiFi.

3. Build seven exciting, instruction-based home automation projects with Arduino in no time.

Raspberry Pi for Secret Agents
Second Edition

ISBN: 978-1-78439-790-6   Paperback: 206 pages

Turn your Raspberry Pi into your very own secret agent toolbox with this set of exciting projects

1. Turn your Raspberry Pi into a multipurpose secret agent gadget for audio/video surveillance, Wi-Fi exploration, or playing pranks on your friends.

2. Detect an intruder on camera and set off an alarm and also find out what the other computers on your network are up to.

3. Full of fun, practical examples and easy-to-follow recipes, guaranteeing maximum mischief for all skill levels.

BeagleBone for Secret Agents

ISBN: 978-1-78398-604-0   Paperback: 162 pages

Browse anonymously, communicate secretly, and create custom security solutions with open source software, the BeagleBone Black, and cryptographic hardware

1. Interface with cryptographic hardware to add security to your embedded project, securing you from external threats.

2. Use and build applications with trusted anonymity and security software like Tor and GPG to defend your privacy and confidentiality.

3. Work with low level I/O on BeagleBone Black like I2C, GPIO, and serial interfaces to create custom hardware applications.

Please check **www.PacktPub.com** for information on our titles

www.ingramcontent.com/pod-product-compliance
Lightning Source LLC
LaVergne TN
LVHW081344050326
832903LV00024B/1299